STEMD^2 Research & Development Group
Center on Disability Studies
College of Education
University of Hawaiʻi at Mānoa

http://stemd2.com/

 Printed in the United States of America. First published (2019) by the Center on Disability Studies, University of Hawaiʻi at Mānoa, Honolulu, Hawaiʻi. This document is based upon work supported by the Department of Education, Native Hawaiian Education Act Program under award #S362A180011. Any opinions, findings, and conclusions or recommendations expressed in this material are those of the authors and do not necessarily reflect the views of the United States Department of Education. For further information about this document and Ne'epapa Ka Hana 2.0 project, please contact Dr. Kaveh Abhari at abhari@hawaii.edu.

ISBN: 978-0-9983142-7-3
First release, 2019

Ne'epapa Ka Hana Sixth-Grade Mathematics Resources

Let's Go from Mauka to Makai

Teacher's Guide

Project Director	Kaveh Abhari
Content Developers	Robert G. Young Justin S. Toyofuku
Creative Designer	MyLan Tran
Publication Designer	Robert G. Young

Acknowledgments

We would like to thank Kelli Ching, Crystal Yoo, Katy Parsons, and Robyn Rice for advising on middle school mathematics. Thank you Moa Viebke, Nohea Behler, and Robyn Rice for significant help reviewing and editing. Mahalo nui to Moa Viebke and Nohea Behler for major contributions in writing the introductions.

Contents

Introduction to *Let's Go from Mauka to Makai*

Let's Go from Mauka to Makai has been created by Ne'epapa Ka Hana (NKH) for 6th grade students of Hawai'i. The mathematical activities featured in *Let's Go from Mauka to Makai* have been constructed to enrich students mathematical abilities through culturally responsive material and to increase interest and participation in the mathematical classroom.

Through the collaboration between the Ne'epapa Ka Hana (NKH) 2.0 and STEMD2, the mathematical curriculum of *Let's Go from Mauka to Makai* has been made available for online classroom integration. The math activities are offered on a social platform which allows for online collaboration between students and teachers. The platform is free and accessible to users with internet access at www.community.stemd2.com

Ne'epapa Ka Hana (NKH) develop programs and materials that are culturally responsive to Hawai'i's unique diversity – using research-based practices to support ongoing STEM education efforts across the state. *Let's Go from Mauka to Makai* is the 6th addition to NKH's middle-school book series.

The *Let's Go from Mauka to Makai* book is composed by mathematical learning activities that serve as a supplementary mathematics curriculum to enhance inclusive instruction based on problem-based learning strategies and connectivism principles. The math activities incorporate real-world challenges that reflect Hawai'i's unique culture, society, and geography. The goal of these activities is for both kumu and haumāna to collaborate while thinking critically and creatively – thereby, deepening students' understanding, application, and appreciation for mathematical thinking in the context of Hawaiian and island culture.

Let's Go from Mauka to Makai, highlights the Common Core State Standards through the theme of Hawaiian environment and ecology. The curriculum focuses on supporting Hawaiian's natural habitat, evaluating current environmental threats and illustrate how to incorporate mathematical solutions to solve them.

Lesson Planning Structure

Aloha kākou, e komo mai, and welcome to our Ne'epapa Ka Hana (NKH) guide for kumu! In this first chapter, we want to introduce to you the Lesson Planning Structure for developing, and reflecting upon your lessons. The planning suggestions presented provide guidelines for developing a range of meaningful lessons and memorable activities that truly engage your haumāna.

Formative Steps in the NKH Activity Set

Activities within each unit can be used separately and in various order. However, the cumulative activity should come after the different components of a module has been covered. To maximize the benefits of the activities, we suggest the following.

1. Allow the students to attempt the activity individually.
2. After reviewing students' initial solutions, formulate questions to challenge the students to explain their thought process and improve their responses.
3. Next, arrange students to work in small groups to synthesize their understandings of the activity. Four students per group is recommended, as students could also work in pairs within the group.
4. In the same small groups, students need to explain to each other how they come to the initial response and comment on initial responses by comparing them with their own work.
5. The group continues working together to finalize a solution for the given activity set to present to the class.
6. In a whole-class discussion–preferably moderated by themselves–students compare and evaluate the strategies they have seen and used.
7. In the end, you need to summarize the students' discussions and explain one or two possible solutions.
8. Before moving to the next lesson, ask students to individually reflect for 10 minutes on their work and on what they have learned. You may ask your students to write what they learn in a few sentences.

Along with the book, we also offer online activities available on our community platform at https://community.stemd2.com. We encourage teachers to implement the community platform to provide students an online collaboration forum where they can work with other students on the math activities.

To facilitate the group and class discussion, you may need mini whiteboards with markers and erasers to quickly and visibly check individual understanding. This instructional strategy also enhances attention and participation. Some activities need calculators and graph paper. Using a projector and screen to share students' sample responses are highly recommended as well. A few activities in *Let's Go from Mauka to Makai* require materials such as a ruler with centimeters, and a yard stick or measuring tape.

Lesson Structure

Give each student a copy of the appropriate **NKH Activity** or access to the activities on the NKH community platform.

Examples of instructional prompts for individual students who are attempting any given activity include the following:

- Read the questions and try to answer them as carefully as you can.
- Show all your work, so that I can understand your reasoning.
- In addition to trying to solve the problem, I want you to check if you can present your work in a clear and organized fashion.

Students should work individually and without your assistance. Note that you may have to rearrange your students seating arrangements.

Assessing Students' Responses and Giving Feedback

After collecting the students' attempts at a NKH Activity, please take the time to create a few notes about what these samples of students' work reveal about their current levels of understanding and their unique approaches to problem-solving.

Scoring is not recommended during this phase.

It is also important to note that as a kumu, your feedback should summarize students' difficulties as a series of questions either by:

- Writing one or two questions on each students' work;
- Giving each student a printed version of you list of questions and highlight the questions that are more relevant to each individual student; or
- Selecting a few questions that will be of help to the majority of students and sharing them collectively with the whole class (either projected or written on the board) when returning students' initial attempts at the beginning of the NKH lesson.

When providing feedback to your haumāna, please refer to your own professional judgment and the respective needs of your unique instructional setting. That being said, certain common issues do arise across different classrooms and we recommend the following instructional prompts:

Common Issues	**Feedback Examples**
Student has difficulty getting started	• What do you know? • What do you need to find out?
Student omits some given information when solving the problem	Write the given information in your own words.
Student overlooks or misinterprets some constraints	• Can you organize... in a systematic way? • What would make sense to try? Why? • How can you organize your work?
Student makes incorrect assumptions	Will it always be the way you described?

Continued on next page

Common Issues	Feedback Examples
Work is poorly presented	• Could someone unfamiliar with the task easily understand your work? • Have you explained how you arrived at your answer?
Student provides little or no justification	• How did you get your answer? • How could you convince me that...?
Completes the task early	• What would happen if...? • Double check your work • Can you spot any mistakes?

When you are delivering the full NKH Activity Set as a lesson, please make sure that you have allotted approximately 50 minutes for proper lesson delivery and activity execution.

That being said, reviewing a students' first attempt at a NKH Activity should take approximately 5-10 minutes. This review is done individually and generally provided to the entire class, as a collective whole, by either projecting on a screen or writing on the board.

Collaborative Small-Group Work

Once divided into groups, students should take 10-15 minutes to share and reflect upon their individual attempts at the prior, independent NKH Activity.

During this time, you may provide a few additional guiding prompts for your student groups:

- You each have your own individual solution to the task. Now, I want you to share your work with your partner(s). Take turns to explain how you did the task and how you think it could be improved.
- If explanations are unclear, ask questions until everyone in the group understands the individual solutions.

This leads to the next 15-20 minutes where students work together towards a common solution. If you are using the NKH community platform, your joint solution can be posted onto the Forum. Of course, the use of your own professional judgment and any other format achieving this goal will suffice.

During this section, you should take note of the different approaches between groups, the change of direction, dialogue between groups, etc. This effort will help you guide the class discussion wrap-up.

You will support and foster problem-solving skills by asking questions that help your haumāna clarify their thinking, while encouraging students to develop self-regulation as well as error detection skills.

Sharing, discussing, analyzing different approaches (10-20 minutes)

In this section, a whole-class discussion may follow the previous section. Voluntary or randomly select groups to share their strategies that were developed towards a joint solution. It may be important to ask how the students' group solution different from their individual solutions. If your students do not explicitly state their conclusions, you might ask how they checked their work. A conversation could be initiated via the online Forum as well.

Wrap-up (5-10 minutes)

As we wrap-up the final sections, a class discussion might conclude the NKH Activity Set by comparing the advantages and disadvantages of the approaches in the activity. Such discussion may center around shared difficulties or possible shortcuts that students could have developed either together or independently. It is important to recognize students' feelings and attitudes both during and after these activities.

Please note that the timing for these sections and activities range from 40-70 minutes but may vary from classroom to classroom depending on the nature of your needs within your particular instructional setting.

Additional Notes

The given lesson duration are only approximate. Please feel free to spend more or less time on these activities if it suits your classroom better. Lastly the icon indicates where students can use the online learning platform.

Please send us any comments, issues, technical or otherwise, you might have with the content, the format, or the approach.

Common Core State Standards

Common Core State Standard	***Let's Go from Mauka to Makai* Unit**
Ratios & Proportional Relationships	
6.RP.1 Understand the concept of a ratio and use ratio language to describe a ratio relationship between two quantities. *For example, "The ratio of wings to beaks in the bird house at the zoo was 2:1, because for every 2 wings there was 1 beak." "For every vote candidate A received, candidate C received nearly three votes."*	3
6.RP.2 Understand the concept of a unit rate a/b associated with a ratio a:b with b ≠ 0, and use rate language in the context of a ratio relationship. *For example, "This recipe has a ratio of 3 cups of flour to 4 cups of sugar, so there is 3/4 cup of flour for each cup of sugar." "We paid $75 for 15 hamburgers, which is a rate of $5 per hamburger."* (Note: Expectations for unit rates in this grade are limited to non-complex fractions.)	3
6.RP.3 Use ratio and rate reasoning to solve real-world and mathematical problems, e.g., by reasoning about tables of equivalent ratios, tape diagrams, double number line diagrams, or equations.	3
6.RP.3a Make tables of equivalent ratios relating quantities with whole-number measurements, find missing values in the tables, and plot the pairs of values on the coordinate plane. Use tables to compare ratios.	3
6.RP.3b Solve unit rate problems including those involving unit pricing and constant speed. *For example, if it took 7 hours to mow 4 lawns, then at that rate, how many lawns could be mowed in 35 hours? At what rate were lawns being mowed?*	3
6.RP.3c Find a percent of a quantity as a rate per 100 (e.g., 30% of a quantity means 30/100 times the quantity); solve problems involving finding the whole, given a part and the percent.	3

Continued on next page

Common Core State Standard	*Let's Go from Mauka to Makai* Unit
6.RP.3d Use ratio reasoning to convert measurement units; manipulate and transform units appropriately when multiplying or dividing quantities.	3
The Number System	
6.NS.1 Interpret and compute quotients of fractions, and solve word problems involving division of fractions by fractions, e.g., by using visual fraction models and equations to represent the problem. *For example, create a story context for (2/3) ÷ (3/4) and use a visual fraction model to show the quotient; use the relationship between multiplication and division to explain that (2/3) ÷ (3/4) = 8/9 because 3/4 of 8/9 is 2/3. (In general, (a/b) ÷ (c/d) = ad/bc.). How much chocolate will each person get if 3 people share 1/2 lb of chocolate equally? How many 3/4-cup servings are in 2/3 of a cup of yogurt? How wide is a rectangular strip of land with length 3/4 mi and area 1/2 square mi?*	2
6.NS.2 Fluently divide multi-digit numbers using the standard algorithm.	2
6.NS.3 Fluently add, subtract, multiply, and divide multi-digit decimals using the standard algorithm for each operation.	2
6.NS.4 Find the greatest common factor of two whole numbers less than or equal to 100 and the least common multiple of two whole numbers less than or equal to 12. Use the distributive property to express a sum of two whole numbers 1-100 with a common factor as a multiple of a sum of two whole numbers with no common factor. *For example, express 36 + 8 as 4 (9 + 2).*	1
6.NS.5 Understand that positive and negative numbers are used together to describe quantities having opposite directions or values (e.g., temperature above/below zero, elevation above/below sea level, debits/credits, positive/negative electric charge); use positive and negative numbers to represent quantities in real-world contexts, explaining the meaning of 0 in each situation.	1
6.NS.6 Understand a rational number as a point on the number line. Extend number line diagrams and coordinate axes familiar from previous grades to represent points on the line and in the plane with negative number coordinates.	1, 5
6.NS.6a Recognize opposite signs of numbers as indicating locations on opposite sides of 0 on the number line; recognize that the opposite of the opposite of a number is the number itself, e.g., −(−3) = 3, and that 0 is its own opposite.	1

Continued on next page

Common Core State Standard	*Let's Go from Mauka to Makai* Unit
6.NS.6b Understand signs of numbers in ordered pairs as indicating locations in quadrants of the coordinate plane; recognize that when two ordered pairs differ only by signs, the locations of the points are related by reflections across one or both axes.	5, 6
6.NS.6c Find and position integers and other rational numbers on a horizontal or vertical number line diagram; find and position pairs of integers and other rational numbers on a coordinate plane.	1, 5
6.NS.7 Understand ordering and absolute value of rational numbers.	1
6.NS.7a Interpret statements of inequality as statements about the relative position of two numbers on a number line diagram. *For example, interpret −3 > −7 as a statement that −3 is located to the right of −7 on a number line oriented from left to right.*	1
6.NS.7b Write, interpret, and explain statements of order for rational numbers in real-world contexts. *For example, write −3° C > −7° C to express the fact that −3° C is warmer than −7° C.*	1
6.NS.7c Understand the absolute value of a rational number as its distance from 0 on the number line; interpret absolute value as magnitude for a positive or negative quantity in a real-world situation. *For example, for an account balance of −30 dollars, write |−30| = 30 to describe the size of the debt in dollars.*	1
6.NS.7d Distinguish comparisons of absolute value from statements about order. *For example, recognize that an account balance less than −30 dollars represents a debt greater than $30.*	1
6.NS.8 Solve real-world and mathematical problems by graphing points in all four quadrants of the coordinate plane. Include use of coordinates and absolute value to find distances between points with the same first coordinate or the same second coordinate.	5, 6
Equations & Expressions	
6.EE.1 Write and evaluate numerical expressions involving whole-number exponents.	4
6.EE.2 Write, read, and evaluate expressions in which letters stand for numbers.	4

Continued on next page

Common Core State Standard	***Let's Go from Mauka to Makai* Unit**
6.EE.2a Write expressions that record operations with numbers and with letters standing for numbers. *For example, express the calculation "Subtract y from 5" as 5 – y.*	4
6.EE.2b Identify parts of an expression using mathematical terms (sum, term, product, factor, quotient, coefficient); view one or more parts of an expression as a single entity. *For example, describe the expression 2 (8 + 7) as a product of two factors; view (8 + 7) as both a single entity and a sum of two terms.*	4
6.EE.2c Evaluate expressions at specific values of their variables. Include expressions that arise from formulas used in real-world problems. Perform arithmetic operations, including those involving whole-number exponents, in the conventional order when there are no parentheses to specify a particular order (Order of Operations). *For example, use the formulas $V = s^3$ and $A = 6s^2$ to find the volume and surface area of a cube with sides of length s = 1/2.*	4
6.EE.3 Apply the properties of operations as strategies to generate equivalent expressions. *For example, apply the distributive property to the expression 3(2 + x) to produce the equivalent expression 6 + 3x; apply properties of operations to y + y + y to produce the equivalent expression 3y.*	4
6.EE.4 Identify when two expressions are equivalent (i.e., when the two expressions name the same number regardless of which value is substituted into them). *For example, the expressions y + y + y and 3y are equivalent because they name the same number regardless of which number y stands for.*	4
6.EE.5 Understand solving an equation or inequality as a process of answering a question: which values from a specified set, if any, make the equation or inequality true? Use substitution to determine whether a given number in a specified set makes an equation or inequality true.	5
6.EE.6 Use variables to represent numbers and write expressions when solving a real-world or mathematical problem; understand that a variable can represent an unknown number, or, depending on the purpose at hand, any number in a specified set.	4, 5
6.EE.7 Solve real-world and mathematical problems by writing and solving equations of the form x + p = q and px = q for cases in which p, q and x are all nonnegative rational numbers.	5, 6

Continued on next page

Common Core State Standard	*Let's Go from Mauka to Makai* Unit
6.EE.8 Write an inequality of the form x > c or x < c to represent a constraint or condition in a real-world or mathematical problem. Recognize that inequalities of the form x > c or x < c have infinitely many solutions; represent solutions of such inequalities on number line diagrams.	5
6.EE.9 Use variables to represent two quantities in a real-world problem that change in relationship to one another; write an equation to express one quantity, thought of as the dependent variable, in terms of the other quantity, thought of as the independent variable. Analyze the relationship between the dependent and independent variables using graphs and tables, and relate these to the equation. *For example, in a problem involving motion at constant speed, list and graph ordered pairs of distances and times, and write the equation d = 65t to represent the relationship between distance and time.*	5
Geometry	
6.G.1 Find area of right triangles, other triangles, special quadrilaterals, and polygons by composing into rectangles or decomposing into triangles and other shapes; apply these techniques in the context of solving real-world and mathematical problems.	5, 6
6.G.2 Find the volume of a right rectangular prism with fractional edge lengths by packing it with unit cubes of the appropriate unit fraction edge lengths, and show that the volume is the same as would be found by multiplying the edge lengths of the prism. Apply the formulas V = l w h and V = b h to find volumes of right rectangular prisms with fractional edge lengths in the context of solving real-world and mathematical problems.	6
6.G.3 Draw polygons in the coordinate plane given coordinates for the vertices; use coordinates to find the length of a side joining points with the same first coordinate or the same second coordinate. Apply these techniques in the context of solving real-world and mathematical problems.	6
6.G.4 Represent three-dimensional figures using nets made up of rectangles and triangles, and use the nets to find the surface area of these figures. Apply these techniques in the context of solving real-world and mathematical problems.	6

Continued on next page

Common Core State Standard	***Let's Go from Mauka to Makai* Unit**
Statistics & Probability	
6.SP.1 Recognize a statistical question as one that anticipates variability in the data related to the question and accounts for it in the answers. *For example, "How old am I?" is not a statistical question, but "How old are the students in my school?" is a statistical question because one anticipates variability in students' ages.*	7
6.SP.2 Understand that a set of data collected to answer a statistical question has a distribution which can be described by its center, spread, and overall shape.	7
6.SP.3 Recognize that a measure of center for a numerical data set summarizes all of its values using a single number, while a measure of variation describes how its values vary using a single number.	7
6.SP.4 Display numerical data in plots on a number line, including dot plots, histograms, and box plots.	7
6.SP.5 Summarize numerical data sets in relation to their context, such as by:	7
6.SP.5a Reporting the number of observations.	7
6.SP.5b Describing the nature of the attribute under investigation, including how it was measured and its units of measurement.	7
6.SP.5c Giving quantitative measures of center (median and/or mean) and variability (interquartile range and/or mean absolute deviation), as well as describing any overall pattern and any striking deviations from the overall pattern with reference to the context in which the data were gathered.	7
6.SP.5d Relating the choice of measures of center and variability to the shape of the data distribution and the context in which the data were gathered.	7

Unit 1: Numbers

In this unit, we'll learn how to use positive and negative integers, fractions, and decimals to describe situations through Hawai'i's vast biodiversity and unique ecosystem. There are four activities in this unit. *Module 1* involves working with integers to find species at different altitudes on the Islands of Hawai'i, *Module 2* focuses on how to use factors and multiples to support our native environment, and *Module 3* explores the rain seasons of Hawai'i with the help of rational numbers. The final activity is cumulative and incorporates concepts from each of the previous activities in this unit.

Common Core State Standards

Common Core State Standard	Module 1	Module 2	Module 3	Unit 1
6.NS.4 Find the greatest common factor of two whole numbers less than or equal to 100 and the least common multiple of two whole numbers less than or equal to 12. Use the distributive property to express a sum of two whole numbers 1-100 with a common factor as a multiple of a sum of two whole numbers with no common factor. *For example, express 36 + 8 as 4 (9 + 2).*		X		X
6.NS.5 Understand that positive and negative numbers are used together to describe quantities having opposite directions or values (e.g., temperature above/below zero, elevation above/below sea level, debits/credits, positive/negative electric charge); use positive and negative numbers to represent quantities in real-world contexts, explaining the meaning of 0 in each situation.	X		X	X
6.NS.6 Understand a rational number as a point on the number line. Extend number line diagrams and coordinate axes familiar from previous grades to represent points on the line and in the plane with negative number coordinates.	X		X	X
6.NS.6a Recognize opposite signs of numbers as indicating locations on opposite sides of 0 on the number line; recognize that the opposite of the opposite of a number is the number itself, e.g., −(−3) = 3, and that 0 is its own opposite.	X		X	X
6.NS.6c Find and position integers and other rational numbers on a horizontal or vertical number line diagram; find and position pairs of integers and other rational numbers on a coordinate plane.	X		X	X
6.NS.7 Understand ordering and absolute value of rational numbers.	X		X	
6.NS.7a Interpret statements of inequality as statements about the relative position of two numbers on a number line diagram. *For example, interpret −3 > −7 as a statement that −3 is located to the right of −7 on a number line oriented from left to right.*	X		X	X

Continued on next page

Common Core State Standard	Module 1	Module 2	Module 3	Unit 1
6.NS.7b Write, interpret, and explain statements of order for rational numbers in real-world contexts. *For example, write −3° C > −7° C to express the fact that −3° C is warmer than −7° C.*	X		X	X
6.NS.7c Understand the absolute value of a rational number as its distance from 0 on the number line; interpret absolute value as magnitude for a positive or negative quantity in a real-world situation. *For example, for an account balance of −30 dollars, write |−30| = 30 to describe the size of the debt in dollars.*	X		X	
6.NS.7d Distinguish comparisons of absolute value from statements about order. *For example, recognize that an account balance less than −30 dollars represents a debt greater than $30.*	X		X	X

Module 1: Integers *Activity*

Biodiversity refers to how many different kinds of living organisms are present in an area. For example, a farm with 1000 chickens has less biodiversity than a farm with a chicken, a duck, a cow, and a pig. The Hawaiian Islands are known for having incredible biodiversity. From the bottom of the ocean to the top of the mountains, we can find a large variety of plants, animals, and other types of life.

Let's take a few days to travel the Big Island of Hawai'i. We will write down some of the life we see and the altitude (height) above sea level that we see it.

We will start the week by exploring the mountains. On the top of Mauna Kea we find a rare Wēkiu beetle at an altitude of 3725 meters. At an altitude of 1432 meters on a nearby mountain, we see the state bird of Hawai'i, the Nēnē goose. On the way down, we see an 'ōhi'a lehua plant with beautiful flowers at an altitude of 1202 meters.

Wēkiu beetle

Nēnē goose

'Ōhi'a lehua

On Friday, we sail out into the ocean. You lower a fishing line down to -642 meters and catch an opakapaka. Your friend lowers a line as deep as she can. Her line goes down 1432 meters below sea level, and she pulls up a skinny fish, with a huge mouth, called a gulper eel.

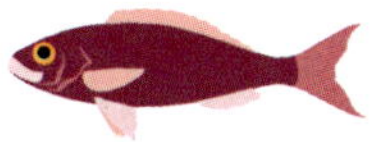

Opakapaka

Gulper eel

The following weekend, we visit a white sandy beach. While swimming offshore, we see a tiny krill at an altitude of about -1 meter (or 1 meter below sea level). At sea level, we spot a Portuguese man o' war floating around. We swim around it and carefully climb 1 meter up a few slippery rocks to collect some delicious opihi.

Krill

Portuguese man o' war

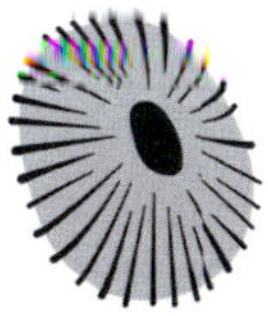

Opihi

1. Use the following table to write down the names of the plants and animals that we saw and the altitude that we saw them. Be sure to use negative numbers for animals that were found below sea level.

Plant/animal name	Altitude spotted (meters)
Wēkiu beetle	3725
Nēnē goose	1432
ʻŌhiʻa lehua	1202
Opakapaka	−642
Gulper eel	−1432
Krill	−1
Portuguese man oʼ war	0
Opihi	1

2. Write down the name of the plants or animals in a list from **lowest altitude to highest altitude**.

(Lowest altitude) (Highest altitude)

Gulper eel Opakapaka Krill Portuguese man oʼ war Opihi ʻŌhiʻa lehua Nēnē goose Wēkiu beetle

3. Write down the name of the plants or animals from **closest to furthest from sea level** (0 meters). If two plants/animals have the same distance from sea level, then write them next to each other.

(Closest to sea level) (Furthest from sea level)

Portuguese man oʼ war	Opihi	Opakapaka	ʻŌhiʻa lehua	Nēnē goose	Wēkiu beetle
	Krill			Gulper eel	

4. With some practice, many people are able to dive down to **10** meters below sea level or hike up to an altitude of 1500 meters without much effort. Let x be the altitudes that we can go to.

   (a) If we cannot go any lower than 10 meters below sea level, then what are the altitudes that we **can** go to?

   (Circle one): $x \geq 10$ $x \leq 10$ $x \geq -10$ (circled) $x \leq -10$

   (b) If we cannot go any higher than an altitude of 1500 meters, then what are the altitudes that we **can** go to?

   (Circle one): $x \geq 1500$ $x \leq 1500$ (circled) $x \geq -1500$ $x \leq -1500$

   (c) List the animals that could have been spotted just by diving or hiking.

   Looking at part 2, we know that the krill, Portuguese man o' war, opihi, and the nēnē could have been spotted between the altitudes of -10 and 1500 meters.

5. Which of these plants and animals have you seen in real life? Where did you see them?

6. What is the most interesting plant or animal that you have seen in Hawai'i? What was so interesting about it?

Module 2: Factors and Multiples *Activity*

To improve the condition of our environment, the government has decided to protect a large slice of land called an ahupua'a. All ahupua'a are made up of two parts, the mauka and the makai. The mauka part is close to the mountains and the makai part is close to the sea. We don't know the exact measurements of the protected ahupua'a, but we do know that the **mauka part has an area of 128 square kilometers**, and the **makai part has an area of 96 square kilometers**. We also know that the ahupua'a and its parts are shaped like rectangles.

For example, if the protected ahupua'a is 8 kilometers wide, then the mauka part must be 16 kilometers long and the makai part must be 12 kilometers long.

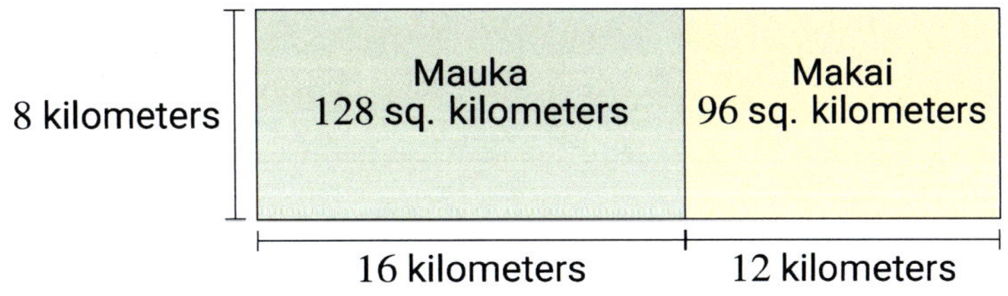

For another example, if the protected ahupua'a is only 4 kilometers wide, then the mauka part must be 32 kilometers long and the makai part must be 24 kilometers long.

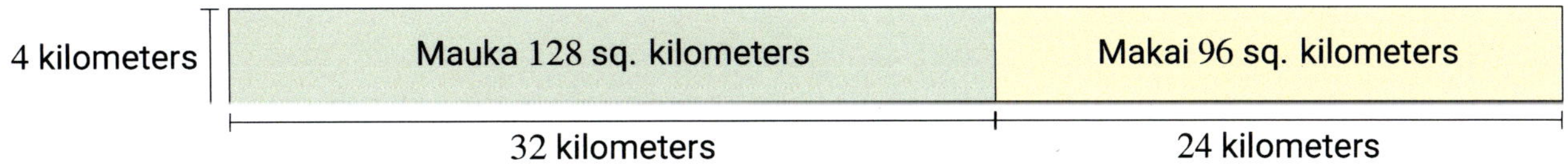

1. Suppose that the width is a **whole number**, draw at least two more possible rectangles to represent the protected area. Be sure to label the sides of your drawings just like in the example above.

The following combinations are possible.

Ahupua'a width (km)	Mauka length (km)	Makai length (km)	
1	128	96	
2	64	48	
4	32	24	(shown in example)
8	16	12	(shown in example)
16	8	6	
32	4	3	

2. If the width is a **whole number**, what is the largest width that the rectangle can be?

The maximum width is also the greatest common factor of the mauka area of 128 and the makai area of 96. The prime factorization of 128 is $2\times2\times2\times2\times2\times2\times2$ (or 2^7 in exponential form) and the prime factorization of 96 is $2\times2\times2\times2\times2\times3$ (or $2^5\times3$ in exponential form). So the greatest common factor is $2\times2\times2\times2\times2$ (or 2^5) and thus the maximum width is 32 kilometers. We could've also found this by listing the factors of 128 and the 96.

Factors of 128: 1, 2, 4, 8, 16, <u>32</u>, 64, 128
Factors of 96: 1, 2, 3, 4, 6, 8, 12, 16, 24, <u>32</u>, 48, 96 <u>Greatest Common Factor</u>

The government is asking for your help in taking care of this area by visiting it regularly. They want you to visit the mauka part **once every 8 weeks** and let them know if that area is doing well. They also want you the visit the makai part **once every 3 weeks** and let them know if it is doing well. This week, you will check on both the mauka and makai areas.

3. Sometimes you will have to visit both parts in the same week. **How often** does this happen?

   To find out when "once every 8 weeks" overlaps with "once every 3 weeks," we need to find the least common multiple (LCM) of 8 and 3. We can see that the LCM of 8 and 3 is 24 so we'll have to visit both the mauka and makai parts once every 24 weeks.

   Multiples of 3: 3, 6, 9, 12, 15, 18, 21, 24
   Multiples of 8: 8, 16, 24 Least Common Multiple

4. If you needed to take care of the ahupua'a for the next 80 weeks, during which weeks would you visit both parts? **How many times** will you need to visit both parts in the same week? Be sure to count this week as the first time.

   If we visited both parts this week then we will have to visit again 24, 48, and 72 weeks from now. So we will have visited both parts 4 times, counting this week as the first.

5. If you visited these protected areas, what would you look for to determine if the area is doing okay? Why would you look for these specific things?

Module 3: Rational Numbers *Activity*

Mount Wai'ale'ale on Kaua'i is one of the rainiest places on Earth. Normally, rainfall is measured in inches, but since it rains so much on Mount Wai'ale'ale, we can measure it in feet.

Mount Wai'ale'ale

We want to compare the rainfall on Mt. Wai'ale'ale with another rainy place called Mawsynram ("mah-sin-rem") in India by calculating the difference. Let's take the rainfall from Mt. Wai'ale'ale and subtract the rainfall from Mawsynram. Here is the data.

	Rainfall (feet)		
Season	Mt. Wai'ale'ale	Mawsynram	Difference
Spring	$8\frac{1}{2}$	$6\frac{1}{5}$	$2\frac{3}{10}$
Summer	7.3	21.5	−14.2
Fall	$7\frac{3}{5}$	$4\frac{2}{5}$	$3\frac{1}{5}$
Winter	7.05	0.2	6.85

1. In the table above, are there any pairs of numbers that are exactly opposite? If so, which pairs?

   No, none of the numbers are exactly opposite. There is only one negative number, -14.2, but its opposite is not in the table.

2. For each season, plot the "difference" column on the number line. (Do not plot any of the other numbers on the table.)

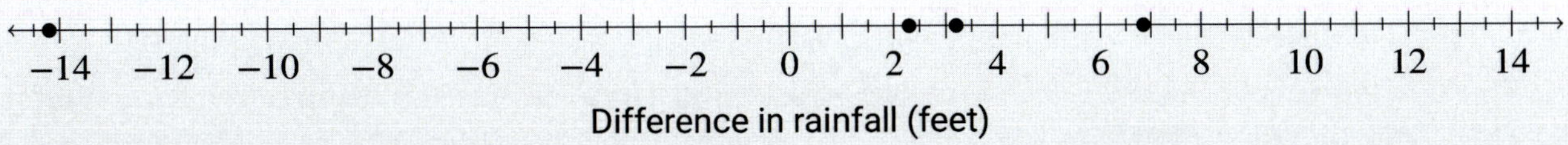

3. When the difference in rain is 0, both Mt. Wai'ale'ale and Mawsynram have the same amount of rainfall. What about when the difference is negative or positive?

   Mt. Wai'ale'ale / Mawsynram (circled) has more rainfall when the difference is **negative**.

   (circle one)

   Mt. Wai'ale'ale (circled) / Mawsynram has more rainfall when the difference is **positive**.

   (circle one)

4. Which season has the largest difference between rainfall in Mount Wai'ale'ale and Mawsynram? How can you tell?

   The rainfall in Mount Wai'ale'ale is most different from Mawsynram during the summer season. You can tell because the the absolute value of the differences for that season is the greatest. For all other seasons, the absolute value of the differences are less than 7. But in the summer, the difference is greater than 14.

5. Which season has the smallest difference between rainfall in Mount Wai'ale'ale and Mawsynram? How can you tell?

   The rainfall in Mount Wai'ale'ale is most similar to Mawsynram during the spring season. You can tell because the the absolute value of the differences for that season is the least, and "least different" is the same as "most similar." It is important to note, however, that the difference is still over 2 feet. That's a lot of rain!

6. Let's compare the "difference" in the summer with the seasons that are not summer.

   (a) Write down the difference in rainfall in the summer.

   The difference in summer is -14.2 feet.

   (b) Add up all of the differences from **spring**, **fall**, and **winter** to find the difference of the seasons that are not summer.

   $$\begin{aligned}\text{Spring + Fall + Winter} &= 2\tfrac{3}{10} + 3\tfrac{1}{5} + 6.85 \\ &= 2.3 + 3.2 + 6.85 \\ &= 12.35\end{aligned}$$

   Total from spring, fall, and winter is 12.35 feet.

7. In one year, does it rain more in Mount Wai'ale'ale or Mawsynram? How can you use part 6 to help you answer this question?

   We see in part 6 that the absolute value of difference in summer is greater than in the seasons that are not summer. I.e. the difference is "more negative that it is positive," showing that it rains a lot more in Mawsynram than on Mount Wai'ale'ale. We can also reach this conclusion by showing that the sum of the two answers in part 6 is negative.

8. Of these two places, which one would you consider to have "the most unusual" weather and why?

   This question is open-ended. Although it rains more in Mount Wai'ale'ale than Mawsynram throughout the year, it rains much more in Mawsynran in the summer which is enough to give it more rain than Mount Wai'ale'ale for the year. Some students may feel that having year-long rains (Mount Wai'ale'ale) is unusual. Others may feel that a summer of massive rainfall (Mawsynram) is unusual.

Unit 1: *Cumulative Activity*

Ancient Hawai'i used to be full of strong koa trees and fragrant 'iliahi (sandalwood) trees, but centuries of deforestation have made most of these trees and the animals that depend on them disappear. Now volunteers are working to bring the trees back.

Koa sapling

'Iliahi sapling

Kahā and Kekai are two gardeners who specialize in growing saplings (baby trees) of native plants. Kahā plans to donate 180 koa saplings, and Kekai will donate 120 'iliahi saplings. They will put them into packages and give them to volunteers who will plant and take care of them. Each package will contain the same mix of saplings, and no saplings will be left over once the packages are filled.

1. Kahā and Kekai are trying to decide how many koa and 'iliahi saplings to put in each package. For example, if each package has 18 koa and 12 'iliahi, then they can make 10 packages.

   Work with a partner to list **other possible ways** to package the koa and 'iliahi.

   To answer this, we need to consider the common factors of 180 and 120. These common factors tell us the number of packages that can be made. Besides 10, which was used in the given example, the other factors are 2, 3, 4, 5, 6, 12, 15, 20, 30, and 60. There is also the factor 1, which is commonly ignored but has value in this context.

Koa per package	**18**	Koa per package	180	Koa per package	90
'Iliahi per package	**12**	'Iliahi per package	120	'Iliahi per package	60
Number of packages	**10**	Number of packages	1	Number of packages	2
Koa per package	60	Koa per package	45	Koa per package	36
'Iliahi per package	40	'Iliahi per package	30	'Iliahi per package	24
Number of packages	3	Number of packages	4	Number of packages	5
Koa per package	30	Koa per package	15	Koa per package	12
'Iliahi per package	20	'Iliahi per package	10	'Iliahi per package	8
Number of packages	6	Number of packages	12	Number of packages	15
Koa per package	9	Koa per package	6	Koa per package	3
'Iliahi per package	6	'Iliahi per package	4	'Iliahi per package	2
Number of packages	20	Number of packages	30	Number of packages	60

2. What is the maximum number of packages Kahā and Kekai can make?

   The maximum number of packages is also the greatest common factor. So in our activity, the greatest number of packages that can be made is 60.

After receiving the donations, the volunteers begin to plant their saplings. Over the next few weeks, many trees will be planted but some will die. Below is a table that shows how the number of trees have changed since the saplings were planted. For example, "15" means that there are 15 more trees now than before the saplings were planted. "−5" means that there are 5 less trees now than before the saplings were planted.

Week	Change in number of trees
0	0
1	3
2	−2
3	−6
4	−5
5	1
6	6

3. For each week, plot the "change in number of trees" on the number line. Make sure to label the number line.

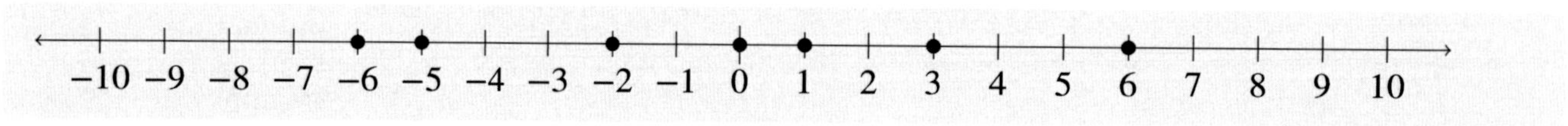

4. Determine whether the following statements are **true or false**.

   (a) There were more trees in week 3 than week 2. True or False

   (b) There were more trees in week 4 than week 3. True or False

   (c) There were more trees in week 3 than week 6. True or False

   (d) There were more trees in week 5 than week 4. True or False

   (e) Two of the numbers in "change in number of trees" column are opposite. True or False

5. Restoring a forest doesn't always work as planned. A new sapling has to fight for sun, water, and nutrients while facing bad weather, pollution, and pests. If you wanted to learn how to grow a plant, what kind of plant would you choose and why? Please share your thoughts with your partner or in the online comment section.

Unit 2: Number Operations

In this unit, we'll learn how to use positive integers, fractions, and decimals to solve problems through exploration of invasive species and evaluating Hawai'i's ecosystem. There are three activities in this unit. *Module 4* involves tracking and evaluating the coconut beetle through the use of operations with fractions. *Module 5* focuses on operations with decimals by evaluating the rain fall on a ahupua'a. The final activity is cumulative and incorporates concepts from each of the previous activities in this unit.

Common Core State Standards

Common Core State Standard	Module 4	Module 5	Unit 2
6.NS.1 Interpret and compute quotients of fractions, and solve word problems involving division of fractions by fractions, e.g., by using visual fraction models and equations to represent the problem. *For example, create a story context for (2/3) ÷ (3/4) and use a visual fraction model to show the quotient; use the relationship between multiplication and division to explain that (2/3) ÷ (3/4) = 8/9 because 3/4 of 8/9 is 2/3. (In general, (a/b) ÷ (c/d) = ad/bc.). How much chocolate will each person get if 3 people share 1/2 lb of chocolate equally? How many 3/4-cup servings are in 2/3 of a cup of yogurt? How wide is a rectangular strip of land with length 3/4 mi and area 1/2 square mi?*	X		X
6.NS.2 Fluently divide multi-digit numbers using the standard algorithm.		X	X
6.NS.3 Fluently add, subtract, multiply, and divide multi-digit decimals using the standard algorithm for each operation.		X	X
6.NS.4 Find the greatest common factor of two whole numbers less than or equal to 100 and the least common multiple of two whole numbers less than or equal to 12. Use the distributive property to express a sum of two whole numbers 1-100 with a common factor as a multiple of a sum of two whole numbers with no common factor. *For example, express 36 + 8 as 4 (9 + 2).*	X		X

Module 4: Operations with Fractions *Activity*

A $13\frac{1}{2}$ acre forest has been attacked by invasive coconut rhinoceros beetles. A team of 8 researchers have decided to spread out and check on the health of the forest.

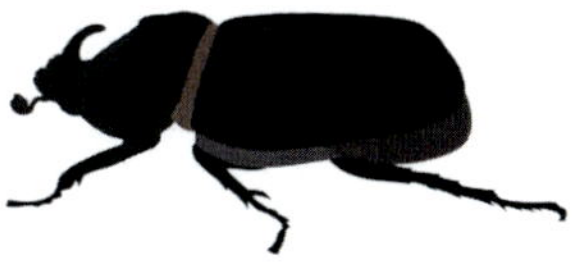

Coconut rhinoceros beetle

1. Each of the 8 researchers have been assigned a equal share of the $13\frac{1}{2}$ acres of forest. This means that they have to watch $1\frac{11}{16}$ or $\frac{27}{16}$ acres of forest each. How did they come up with this number? Please show all of your work below.

   They had to divide the total by 8.

   $$13\frac{1}{2} \div 8 = \frac{27}{2} \div 8 = \frac{27}{2} \times \frac{1}{8} = \frac{27}{16}$$

Sometimes, we have different ways to describe the same thing. For example, when we say that "we lost half of our money," we can describe this by taking the amount of money we had and **dividing** by 2. We can also take the money we had and **multiply** by $\frac{1}{2}$. It's the same thing.

2. After watching the forests for several years, each researcher saw that their part has gotten smaller. They wrote down a **number** that describes how much the forest has changed, but they didn't write down whether you are supposed to **multiply or divide** by that number. Write down the missing sign (× or ÷) that makes the forest size ($\frac{27}{16}$) **smaller**. Find the size of the smaller forest part. Give your answer as a mixed number.

   (a) Researcher 1: $\frac{27}{16} \div \frac{4}{3} = 1\frac{17}{64}$

   $$\frac{27}{16} \div \frac{4}{3} = \frac{27}{16} \times \frac{3}{4} = \frac{81}{64} = 1\frac{17}{64}$$

   (b) Researcher 2: $\frac{27}{16} \times \frac{16}{25} = 1\frac{2}{25}$

   $$\frac{27}{16} \times \frac{16}{25} = \frac{27}{1} \times \frac{1}{25} = \frac{27}{25} = 1\frac{2}{25}$$

   (c) Researcher 3: $\frac{27}{16} \times \frac{8}{15} = \frac{9}{10}$

   $$\frac{27}{16} \times \frac{8}{15} = \frac{9}{2} \times \frac{1}{5} = \frac{9}{10}$$

(d) Researcher 4: $\frac{27}{16} \textcircled{\div} \frac{27}{20} = \frac{5}{4}$

$$\begin{aligned}\frac{27}{16} \div \frac{27}{20} &= \frac{27}{16} \times \frac{20}{27}\\ &= \frac{1}{4} \times \frac{5}{1}\\ &= \frac{5}{4}\end{aligned}$$

(e) Researcher 5: $\frac{27}{16} \textcircled{\div} \frac{18}{5} = \frac{15}{32}$

$$\begin{aligned}\frac{27}{16} \div \frac{18}{5} &= \frac{27}{16} \times \frac{5}{18}\\ &= \frac{3}{16} \times \frac{5}{2}\\ &= \frac{15}{32}\end{aligned}$$

(f) Researcher 6: $\frac{27}{16} \textcircled{\div} \frac{25}{24} = 1\frac{31}{50}$

$$\begin{aligned}\frac{27}{16} \div \frac{25}{24} &= \frac{27}{16} \times \frac{24}{25}\\ &= \frac{27}{2} \times \frac{3}{25}\\ &= \frac{81}{50}\\ &= 1\frac{31}{50}\end{aligned}$$

(g) Researcher 7: $\frac{27}{16} \textcircled{\times} \frac{16}{21} = 1\frac{2}{7}$

$$\begin{aligned}\frac{27}{16} \times \frac{16}{21} &= \frac{9}{1} \times \frac{1}{7}\\ &= \frac{9}{7}\\ &= 1\frac{2}{7}\end{aligned}$$

(h) Researcher 8: $\frac{27}{16} \textcircled{\div} \frac{21}{20} = 1\frac{17}{28}$

$$\begin{aligned}\frac{27}{16} \div \frac{21}{20} &= \frac{27}{16} \times \frac{20}{21}\\ &= \frac{9}{4} \times \frac{5}{7}\\ &= \frac{45}{28}\\ &= 1\frac{17}{28}\end{aligned}$$

Module 5: Operations with Decimals *Activity*

If you have taken a trip around a Hawaiian island, you may have noticed that the weather changes between different ahupua'a. Even neighboring ahupua'a can have very different amounts of wind, rain, and even sunlight. Let's look at three different ahupua'a on a day that has been raining on and off. We recorded the time of each rain shower and the total amount of rain that fell by the end of the day.

Ahupua'a 1

Rain times	1.55 hours 0.12 hours 1 hour
Rain amount	2.1 inches

Ahupua'a 2

Rain times	0.2 hours 5.15 hours 9.8 hours
Rain amount	4.0 inches

Ahupua'a 3

Rain times	0.8 hours 0.5 hours 0.2 hours
Rain amount	2.9 inches

1. On which ahupua'a did the most rain fall?

   Ahupua'a 2 had the most amount of rain at 4.0 inches.

2. For each ahupua'a, calculate the total rain time. Then rank each ahupua'a from **least to greatest** total time spent under rain.

   Ahupua'a 1
   $$\begin{array}{r} 1.55 \\ 0.12 \\ +\ 1.00 \\ \hline 2.67 \end{array}$$

   Ahupua'a 2
   $$\begin{array}{r} 0.20 \\ 5.15 \\ +\ 9.80 \\ \hline 15.15 \end{array}$$

   Ahupua'a 3
   $$\begin{array}{r} 0.8 \\ 0.5 \\ +\ 0.2 \\ \hline 1.5 \end{array}$$

   The ahupua'a rain times from least to greatest occurs in Ahupua'a 3, 1, and 2.

3. The **rain intensity** tells us whether it is raining lightly or whether it is raining heavily (known as "raining cats and dogs"). **Rain intensity** can be found by dividing the total amount of rain by the total rain time. Find the rain intensity for each ahupua'a. Round to two decimal places. Be sure to label your answer in inches per hour.

   Ahupua'a 1
   $2.1 \text{ inches} \div 2.67 \text{ hour} = 0.79 \frac{\text{inches}}{\text{hour}}$

   Ahupua'a 2
   $4.0 \text{ inches} \div 15.15 \text{ hour} = 0.26 \frac{\text{inches}}{\text{hour}}$

   Ahupua'a 3
   $2.9 \text{ inches} \div 1.5 \text{ hour} = 1.93 \frac{\text{inches}}{\text{hour}}$

4. Discuss with a partner which ahupua'a has the lightest rain and which has the heaviest.
5. For each ahupua'a, draw a picture of it raining outside. Make sure it matches your discussion from part 4.

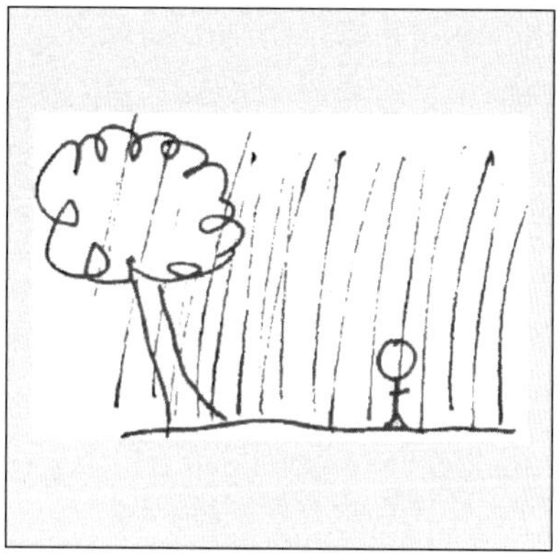

Ahupua'a 1

Ahupua'a 2

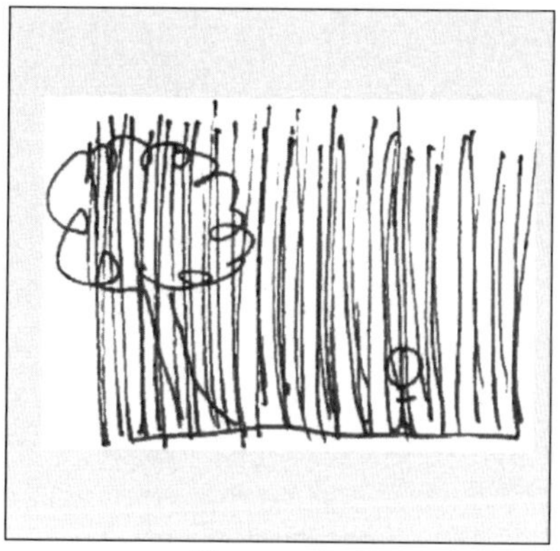

Ahupua'a 3

Let's make sure that the image with the least intense rainfall is in Ahupua'a 2 and the most intense is in Ahupua'a 3.

Unit 2: *Cumulative Activity*

The coconut rhinoceros beetle is an invasive beetle that is spreading and killing both coconut and palm trees. As of 2019, the rhinoceros beetle has been found in the ʻEwa moku of Oʻahu, mostly around Pearl Harbor. (A moku is a group of ahupuaʻa.)

Coconut rhinoceros beetle

You are monitoring a small palm forest on Oʻahu that is under attack by the rhinoceros beetle.

1. The forest originally had a size of 10 acres. (An acre is a little smaller than the size of a football field without the end zones.) A year ago, the beetle killed $\frac{11}{8}$ acres of the forest. How many acres of the forest are still alive? Give your answer as an improper fraction or a mixed number.

$$\begin{aligned} 10 - \frac{11}{8} &= \frac{80}{8} - \frac{11}{8} \\ &= \frac{80-11}{8} \\ &= \frac{69}{8} \end{aligned}$$

After the beetles attacked, $\frac{69}{8}$ acres of forest remain.

2. To learn more about the beetle, scientists went into the remaining parts of the forest (from part 1) and placed a number of beetle traps. They decided to place at least one beetle trap every 3/4 acres of living palm trees. What is the minimum number of traps they would need to place? (Please show your work.)

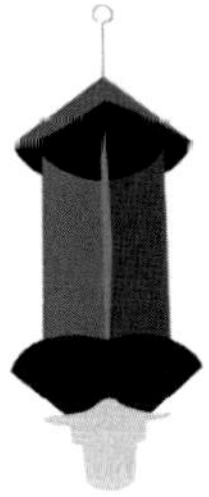

Beetle trap

We need to use division to find out how many $\frac{3}{4}$ are in $\frac{69}{8}$.

$$\begin{aligned} \frac{69}{8} \div \frac{3}{4} &- \frac{69}{8} \times \frac{4}{3} \\ &= \frac{23}{8} \times \frac{4}{1} \\ &= \frac{23}{2} \times \frac{1}{1} \\ &= \frac{23}{2} \\ &= 11\frac{1}{2} \end{aligned}$$

We will need at least 12 traps to cover this area.

3. As the beetles continued to spread, another 3.025 acres of palm trees died in this forest. What is the size of the forest now? Give your answer as a decimal.

$$\begin{aligned} \frac{69}{8} &= 8\frac{5}{8} \\ &= 8 + (5 \div 8) \\ &= 8 + 0.625 \\ &= 8.625 \\ 8.625 - 3.025 &= 5.6 \end{aligned}$$

After the beetles continued to spread, 5.6 acres of forest remain.

4. Your school wants to ask the government to do more to protect these forests. Let's calculate some numbers that can help describe the situation. How many times bigger was the original forest compared to now? Give your answer as a decimal rounded to the nearest tenth.

$10 \div 5.6 = 1.786$

The original forest was about 1.8 times bigger than it is now.

5. A lot of people do not pay attention to environmental issues. If you could get more information about the coconut beetle, what kind of information do you think would make people pay more attention? Share your thoughts in the online comment section. ➚ (Your answer does not have to be a mathy one. =])

6. If you have seen a coconut beetle trap, please take a picture and post it online along with the location of where you saw it! ➚ Or share it with your teacher and your classmates.

Unit 3: Proportionality: Ratio and Rates

In this unit, we'll learn how to use ratios and rates to convert measurements and describe situations through avoiding lava flows and surveying invasive and non-invasive fish populations. There are five activities in this unit. *Module 6* involves representing ratios and rates by evaluating an invasive fish population. *Module 7* has two activities where students guide the keiki to safety by applying ratios and rates to a map in order to avoid Island dangers. *Module 8* explores a variety of fish populations through the use of percents. The final activity is cumulative and incorporates concepts from each of the previous activities in this unit.

Some of the activities in this unit need a ruler with centimeters to complete.

Common Core State Standards

Common Core State Standard	Module 6	Module 7	Module 8	Unit 3
6.RP.1 Understand the concept of a ratio and use ratio language to describe a ratio relationship between two quantities. *For example, "The ratio of wings to beaks in the bird house at the zoo was 2:1, because for every 2 wings there was 1 beak." "For every vote candidate A received, candidate C received nearly three votes."*	X			X
6.RP.2 Understand the concept of a unit rate a/b associated with a ratio a:b with $b \neq 0$, and use rate language in the context of a ratio relationship. *For example, "This recipe has a ratio of 3 cups of flour to 4 cups of sugar, so there is 3/4 cup of flour for each cup of sugar." "We paid $75 for 15 hamburgers, which is a rate of $5 per hamburger."* (Note: Expectations for unit rates in this grade are limited to non-complex fractions.)	X	X		X
6.RP.3 Use ratio and rate reasoning to solve real-world and mathematical problems, e.g., by reasoning about tables of equivalent ratios, tape diagrams, double number line diagrams, or equations.	X	X	X	X
6.RP.3a Make tables of equivalent ratios relating quantities with whole-number measurements, find missing values in the tables, and plot the pairs of values on the coordinate plane. Use tables to compare ratios.		X		
6.RP.3b Solve unit rate problems including those involving unit pricing and constant speed. *For example, if it took 7 hours to mow 4 lawns, then at that rate, how many lawns could be mowed in 35 hours? At what rate were lawns being mowed?*		X		X
6.RP.3c Find a percent of a quantity as a rate per 100 (e.g., 30% of a quantity means 30/100 times the quantity); solve problems involving finding the whole, given a part and the percent.			X	X
6.RP.3d Use ratio reasoning to convert measurement units; manipulate and transform units appropriately when multiplying or dividing quantities.		X	X	X

Module 6: Representing Ratios and Rates *Activity*

When fish or other living things move to new locations, the fish can completely take over its new environment and everything that lives there. When this happens, the fish is called invasive. If the fish was there originally or if it doesn't cause harm, then the fish is called non-invasive. You are at a river in Downtown Honolulu and you are counting the number of invasive and non-invasive fish. You notice that there is a new fish in the river. Let's track the impact of the new fish on the invasive and non-invasive fish populations over the next few years.

Year	Non-invasive fish	Invasive fish
1	150	6
2	138	12
3	126	18
4	114	24
5	102	30
6	90	36
7	78	42
8	66	48
9	54	54
10	42	60

Examples

Non-invasive fish in Hawaii	Invasive fish in Hawaii
 'O'opu (freshwater goby)	 Bass
 'Ama'ama (striped mullet)	 Catfish
 Āholehole (spotted flagtail)	 Cichlid

1. It looks like the non-invasive fish population is dropping at a constant rate, and the invasive fish population is increasing at a constant rate. **What are the rates?** In other words, how much does the number of fish change per year? Make sure to use positive numbers to show that a number is growing and negative numbers to show that a number is shrinking.

   Non-invasive fish: -12 fish per year   Invasive fish: 6 fish per year

2. Complete the previous table for the remaining six years (years 5 to 10).

3. Find the ratio of *non-invasive* to *invasive* fish on the following years. Be sure to reduce your ratios to lowest terms.

   (a) Year 3? The ratio of non-invasive to invasive fish is 126:18 on Year 3, which reduces to 7:1.

   (b) Year 5? The ratio of non-invasive to invasive fish is 102:30 on Year 5, which reduces to 17:5.

4. In what year will the ratio of *non-invasive* to *invasive* fish be 1:1?

In Year 9, the ratio of non-invasive to invasive fish is 54:54, which reduces to 1:1.

5. How many of each type of fish do you expect to count in year 12? Explain.

Non-invasive fish: 18 fish Invasive fish: 72 fish

We can find this by using the numbers from year 10 and the rates from part 1.

The number of non-invasive fish in year 10 was 42 and this number changes by -12 fish per year.
$42 - 12 - 12 = 18$

The number of invasive fish in year 10 was 60 and this number changes by 6 fish per year.
$60 + 6 + 6 = 72$

6. Many plants and animals like kalo (taro), ʻulu (breadfruit), and the monarch butterfly were brought to Hawaiʻi from other places, but they are **not** considered "invasive." Why do you think that is? Can you name other plants or animals that are not from Hawaiʻi and aren't invasive either? Feel free to use the internet or work with a partner if your teacher allows.

Module 7: Applying Ratios and Rates *Activity 1*

For this activity you will need a ruler with centimeters.

A lava flow is coming! You have to help the keiki navigate this ahupua'a to get home safely.

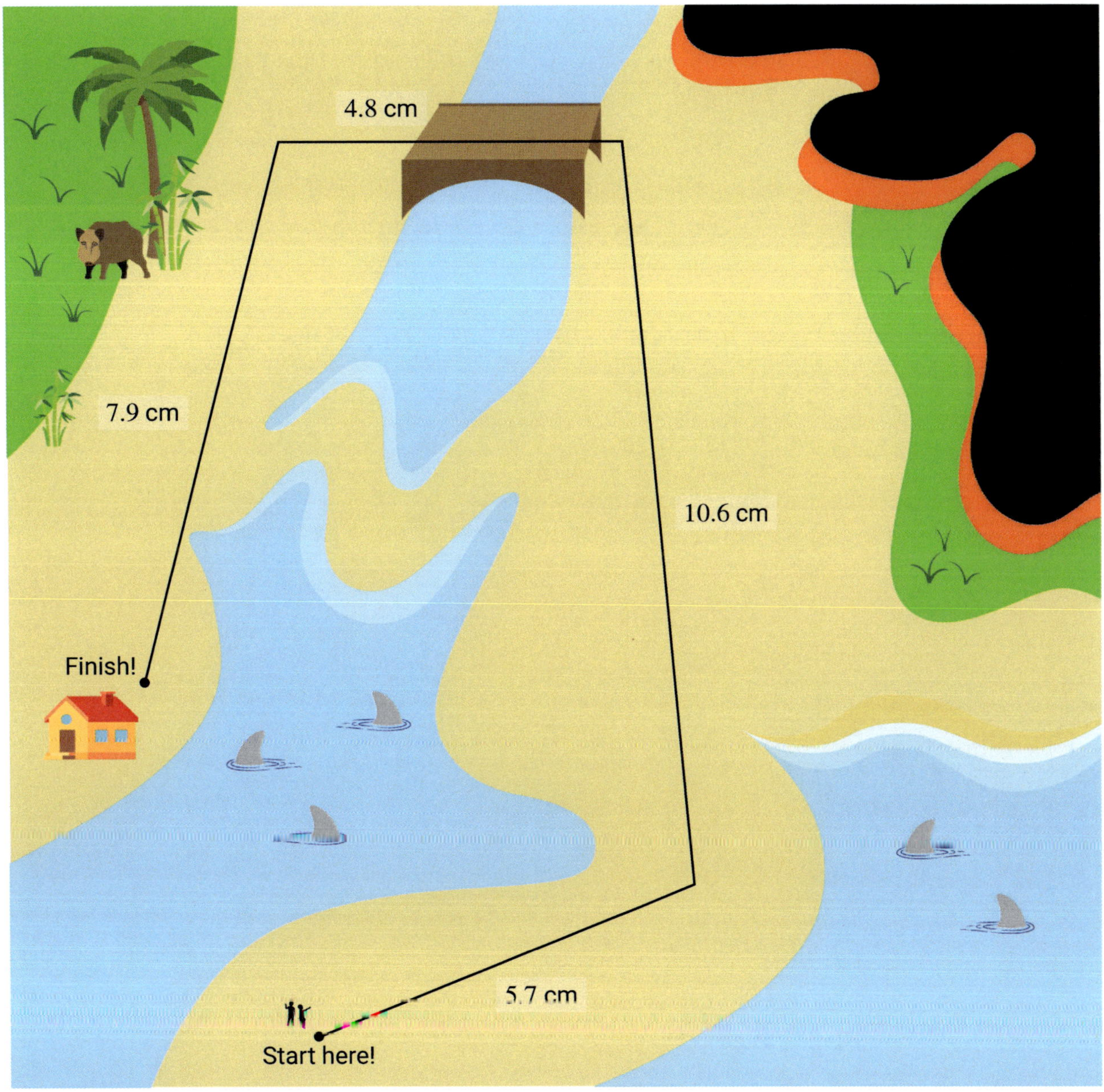

1. Using **straight lines**, draw a path from the **keiki** to the **home**. You may use the bridge or the sandbar to get across the river, but stay out of the water and off of the lava flow. Make sure that there are no breaks in the lines that you draw.

Answers will vary, because there are many different ways to draw the path. Above is just one example.

2. Use your ruler to measure the total length of your lines in centimeters. Round your final total to the nearest centimeter.

Total length: 29 cm

$5.7 + 10.6 + 4.8 + 7.9 = 29.0$

Answers will vary depending on the path drawn. The total length of the path in our example is 29.0 centimeters, which rounds to 29.

3. For this map scale, 40 feet in real life is represented by 5 centimeters on the map. How many feet in real life does each centimeter on the map represent?

$40 \text{ feet} \div 5 \text{ centimeters} = \frac{40 \text{ feet}}{5 \text{ centimeters}} = \frac{40}{5} \frac{\text{feet}}{\text{centimeter}} = 8 \frac{\text{feet}}{\text{centimeter}}$

Each centimeter on the map represents 8 feet in real life.

4. Complete the following table.

Centimeters on the map	1	2	3	4	5	10	11	20
Feet in real life	8	16	24	32	40	80	88	160

5. How many feet was the length of your path?

$29 \text{ cm} \times \frac{8 \text{ feet}}{1 \text{ cm}} = 232 \text{ feet}$

Answers will vary depending on the path drawn. Our path is 29 cm on the map, which represents 232 feet in real life.

Module 7: Applying Ratios and Rates *Activity 2*

This is a follow up of the last activity. Please complete Activity 1 before trying this one! For this activity you will need a ruler with centimeters.

A lava flow is coming! You have to help the keiki navigate this ahupua'a to get home safely.

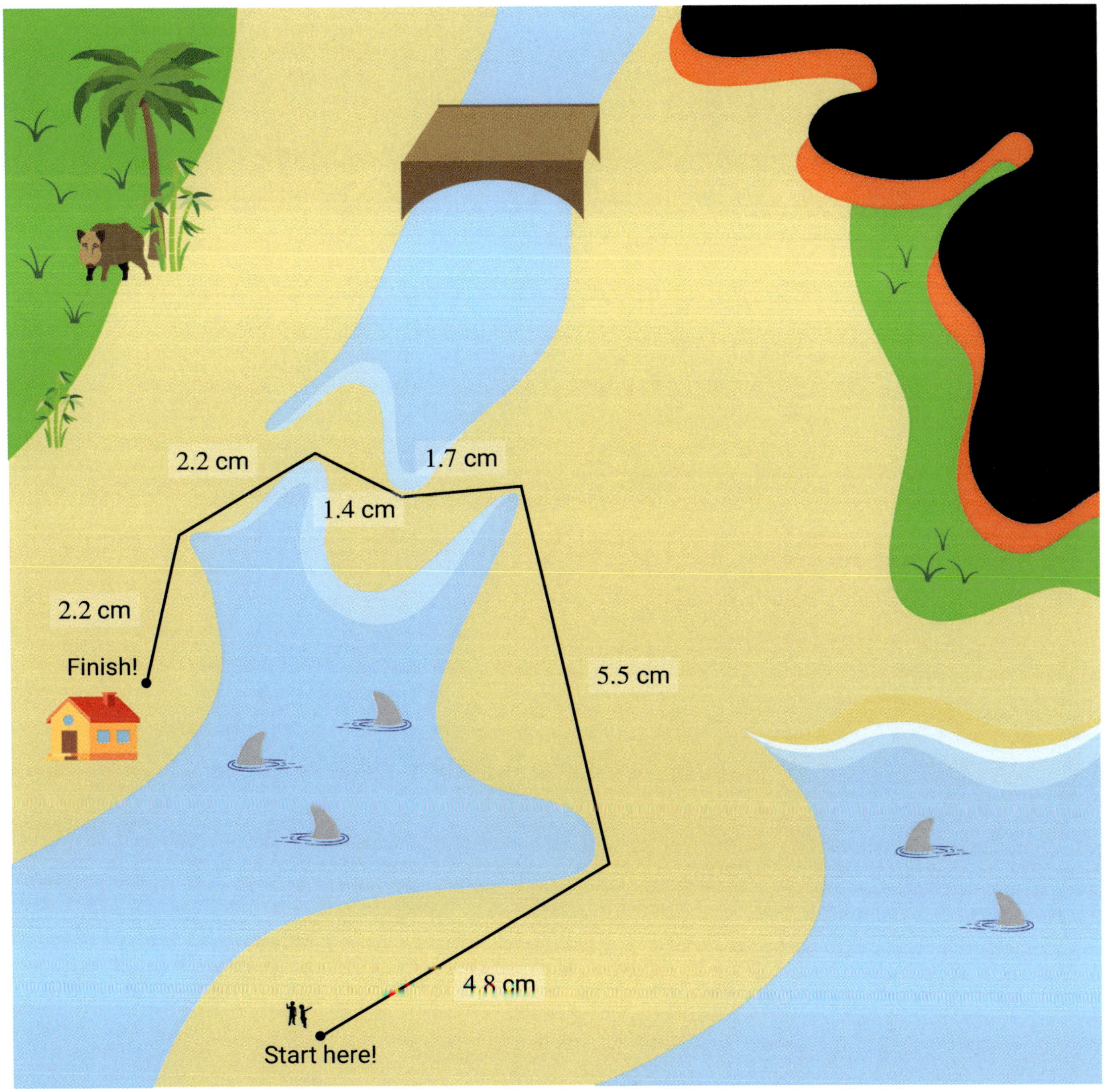

1. This time, let's work with a partner to find the **shortest** path from the keiki to the house. Using **straight lines**, draw a path from the **keiki** to the **home**. You may use the bridge or the sandbar to get across the river, but stay out of the water and off of the lava flow. Make sure that there are no breaks in the lines that you draw.

Answers will vary, but the **shortest** path will only vary slightly from this. The shortest path must cross the sandbar.

2. Find the real distance of your path. Remember, 5 centimeters of your map is 40 feet in real life. Share your solution (map and distance, in feet) and strategy with other groups!

Answers will vary, but this shorter path is about 17.7 cm on the map which rounds to 18.

$$18 \text{ cm} \times \frac{40 \text{ feet}}{5 \text{ cm}} = 18 \text{ cm} \times \frac{8 \text{ feet}}{1 \text{ cm}} = 144 \text{ feet}$$

18 cm on the map is 144 feet in real life.

Module 8: Percents *Activity*

Volunteers are visiting several streams in Honolulu to compare how many fish in the streams are invasive. Unfortunately, each volunteer wrote their numbers in a different formats.

Here is what they found.

Stream	Proportion of invasive fish
Kapālama	19%
Makiki	$\frac{2}{7}$
Mānoa	0.21
Nuʻuanu	9%
Pālolo	$\frac{1}{5}$
Waiʻalae	0.3

Examples

Non-invasive fish in Hawaii	Invasive fish in Hawaii
 ʻOʻopu (freshwater goby)	 Bass
 ʻAmaʻama (striped mullet)	 Catfish
 Āholehole (spotted flagtail)	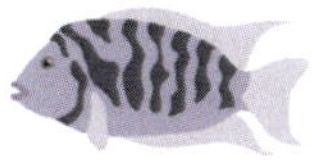 Cichlid

1. One way to get this kind of data is to catch a lot of fish and count how many of the fish are invasive. If the volunteers caught 100 fish in each stream, how many invasive fish would you find in each stream? Round to the nearest whole number if needed. Kapālama is already done for you.

Stream	Number of invasive fish out of 100
Kapālama	19
Makiki	29
Mānoa	21
Nuʻuanu	9
Pālolo	20
Waiʻalae	30

"Percent" means "per hundred." So in a way, we are just converting the proportions to percents. This is most obvious in the Kapālama stream, where 19% of 100 fish is 19.

$$19\% \times 100 \text{ fish} = \frac{19}{100} \times 100 \text{ fish} = 19 \text{ fish}$$

Makiki: $\frac{2}{7} = 2 \div 7 \approx 0.29 = 29\%$

Mānoa: $0.21 = 21\%$

Nuʻuanu: 9%

Pālolo: $\frac{1}{5} = 1 \div 5 = 0.2 = 0.20 = 20\%$

Waiʻalae: $0.3 = 0.30 = 30\%$

2. Rank the streams from the *lowest* proportion of invasive fish to non-invasive fish to *highest*.

(Lowest ratio) (Highest ratio)

Nuʻuanu	Kapālama	Pālolo	Mānoa	Makiki	Waiʻalae

We can use the table in part 1 to help us order the different streams. The original table was difficult to use because the numbers were given in different ways. If we converted the numbers to be all percents, all fractions with common denominators, or all decimals, then we could have used that table as well.

3. Discuss the following questions with a partner. Write your answers and share them with others in your class.

   (a) How do you think the invasive fish got to the rivers in the first place?

   (b) How do you think some of these invasive fish end up taking over a new environment?

Unit 3: *Cumulative Activity*

The pua'a (pig) was introduced to the Hawaiian islands by the first Polynesian explorers. Although this animal is important to the Hawaiian culture, the pua'a can be pretty destructive to the environment. This is mainly because they eat a lot of food and they can eat almost anything. When they are looking for food, they may dig up and chew up native plants. After they eat, they may leave poop that contain seeds of invasive plants.

Pua'a (feral pigs)

The pua'a can often be found eating plants, roots, mushrooms, bugs, and even small mammals. Typically, the diet of a pua'a in your ahupua'a consists of 17 pounds of plants (and fungi) for every 3 pounds of animal meat.

1. Represent this typical diet as a **ratio** of plants (and fungi) to animals.

   17 pounds of plant (and fungi) : 3 pounds of animal

2. What **percent** of this diet would be of plants? Please explain how you got this answer.

   We can find this by dividing the pounds of plant by the total pounds of food. Then, we need to convert this answer to percent.

   $$\frac{17}{17+3} = \frac{17}{20} = 17 \div 20 = 0.85 = 85\%$$

3. What **percent** of this diet would be of animals?

   $$\frac{3}{17+3} = \frac{3}{20} = 17 \div 20 = 0.15 = 15\%$$

4. Suppose that you found a baby pua'a in your yard. You know not to let it go because of the damage that it can do to the environment, so your family decides to keep it as a pet. If you had to prepare 500 pounds of food for this pua'a, how many pounds of plants, and how many pounds of animal meat do you need?

   We can use our answers from the last two problems to help us solve this.

   Plant (and fungi):
   $85\% \times 500 = 0.85 \times 500 = 425$

   Animal:
   $15\% \times 500 = 0.15 \times 500 = 75$

   We need 425 pounds of plants and fungi and 75 pounds of animal meat. This is a lot of food!

5. As we saw before, the pua'a in this ahupua'a eats 17 pounds of plants for every 3 pounds of animal meat. How many pounds of plants does the pua'a eat for one (1) pound of animal meat? Round your answer to the nearest tenth.

 17 pounds of plant to 3 pounds of meat is represented as the fraction $\frac{17}{3}$. $\frac{17}{3}$ is 5.666... in decimal form. This rounds to 5.7 pounds of plant per pound of meat.

 $$\frac{17 \text{ pounds of plant}}{3 \text{ pounds of meat}} = \frac{17}{3}\frac{\text{pounds of plant}}{\text{pound of meat}} = 5.666...\frac{\text{pounds of plant}}{\text{pound of meat}} \approx 5.7\frac{\text{pounds of plant}}{\text{pound of meat}}$$

6. If a baby pua'a is eating 10 pounds of meat, how many pounds of plant should it also eat? Round to the nearest whole number.

 To keep the ratio of 17 pounds of plant to 3 pounds of animal meat, the pua'a should eat 57 pounds of plant with 10 pounds of animal meat.

 $$10 \text{ pounds of meat} \times \frac{5.7 \text{ pounds of plant}}{1 \text{ pound of meat}} = 57 \text{ pounds of plant}$$

7. The diet of a pua'a in one ahupua'a can be different from a pua'a in a different ahupua'a. Why do you think their diets might be different?

Unit 4: Equivalent Expressions

In this unit, we'll learn how to write algebraic expressions in different ways to describe situations through supporting Hawai'i's unique environment by farming. There are three activities in this unit. *Module 9* involves exploring the growth of algae through generating equivalent numerical expressions. *Module 10* focuses on generating equivalent algebraic expressions while evaluating donations to a reforestation project. The final activity is cumulative and incorporates concepts from each of the previous activities in this unit.

Common Core State Standards

Common Core State Standard	Module 9	Module 10	Unit 4
6.EE.1 Write and evaluate numerical expressions involving whole-number exponents.	X		X
6.EE.2 Write, read, and evaluate expressions in which letters stand for numbers.	X	X	X
6.EE.2a Write expressions that record operations with numbers and with letters standing for numbers. *For example, express the calculation "Subtract y from 5" as 5 – y.*			X
6.EE.2b Identify parts of an expression using mathematical terms (sum, term, product, factor, quotient, coefficient); view one or more parts of an expression as a single entity. *For example, describe the expression 2 (8 + 7) as a product of two factors; view (8 + 7) as both a single entity and a sum of two terms.*			X
6.EE.2c Evaluate expressions at specific values of their variables. Include expressions that arise from formulas used in real-world problems. Perform arithmetic operations, including those involving whole-number exponents, in the conventional order when there are no parentheses to specify a particular order (Order of Operations). *For example, use the formulas $V = s^3$ and $A = 6s^2$ to find the volume and surface area of a cube with sides of length s = 1/2.*	X	X	X
6.EE.3 Apply the properties of operations as strategies to generate equivalent expressions. *For example, apply the distributive property to the expression 3(2 + x) to produce the equivalent expression 6 + 3x; apply properties of operations to y + y + y to produce the equivalent expression 3y.*		X	
6.EE.4 Identify when two expressions are equivalent (i.e., when the two expressions name the same number regardless of which value is substituted into them). *For example, the expressions y + y + y and 3y are equivalent because they name the same number regardless of which number y stands for.*		X	X

Continued on next page

Common Core State Standard	Module 9	Module 10	Unit 4
6.EE.6 Use variables to represent numbers and write expressions when solving a real-world or mathematical problem; understand that a variable can represent an unknown number, or, depending on the purpose at hand, any number in a specified set.		X	X

Module 9: Generating Equivalent Numerical Expressions *Activity*

To help the native fish populations, you have decided to raise baby fish at home. First, you need to grow some algae for the fish to eat. You have three types of algae called Type 1, Type 2, and Type 3, and each type grows differently.

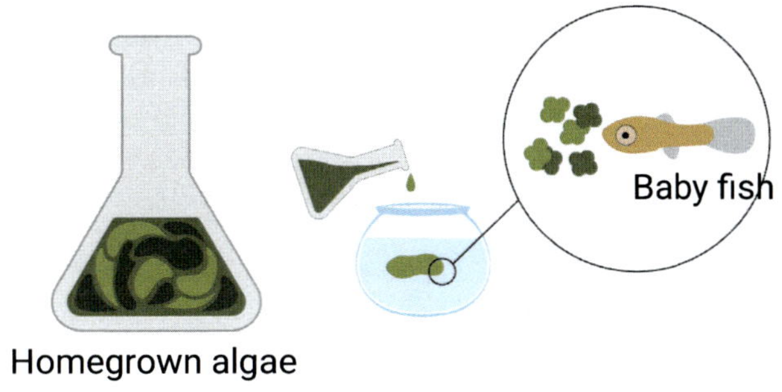

1. The Type 1 algae is tripling its population every day. After two days, its starting population has multiplied by a factor of 9 or 3^2 (in exponent form). After three days, its starting population has multiplied by a factor of 27 or 3^3 (in exponent form). Please fill out the following table.

	Number of times the Type 1 population has multiplied by	
Day	Factor	In exponent form
1	3	3^1
2	9	3^2
3	27	3^3
4	81	3^4
5	243	3^5
6	729	3^6
7	2187	3^7
8	6561	3^8
9	19683	3^9
10	59049	3^{10}

2. The Type 2 algae grows more unpredictably. During the first few days, the algae quintupled (multiplied its population by a factor of 5). Then, it tripled its populations for a few days. Finally, it began to double its population every day. After a total of ten (10) days of growth, the Type 2 algae has multiplied its starting population by a factor of 12960.

   (a) Find the prime factorization of 12960 in exponential form.

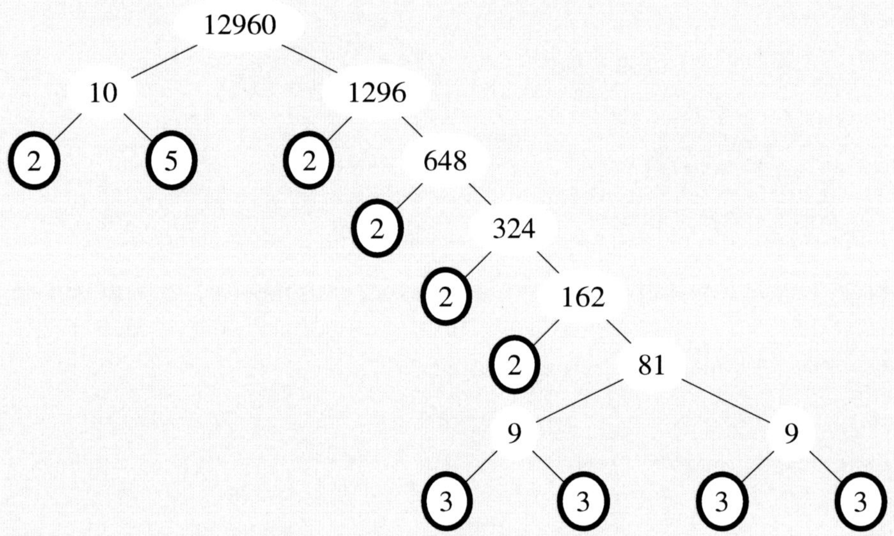

$12960 = 2^5 \times 3^4 \times 5$

   (b) Using your answer from 2a, find out how many days, out of the ten days, the Type 2 algae doubled (×2), tripled (×3), and quintupled (×5) its population.

   Double: 5 Triple: 4 Quintuple: 1

3. After D days, the Type 3 algae has multiplied by the following number.

$$1000 \div \left(1024 \div 2^D - 0.8\right)$$

How many times did the original population of this algae multiply after ten days ($D = 10$)?

$$\begin{aligned}
1000 \div \left(1024 \div 2^D - 0.8\right) &= 1000 \div \left(1024 \div 2^{10} - 0.8\right) \\
&= 1000 \div (1024 \div (2 \times 2 \times 2 \times 2 \times 2 \times 2 \times 2 \times 2 \times 2 \times 2) - 0.8) \\
&= 1000 \div (1024 \div (1024) - 0.8) \\
&= 1000 \div (1 - 0.8) \\
&= 1000 \div (0.2) \\
&= 5000
\end{aligned}$$

After 10 days, the algae has multiplied by 5,000

4. Rank the three types of algae from **slowest** to **fastest** growing after ten (10) days

After 10 days, Type 1 algae has multiplied by 59,049, Type 2 algae has multiplied by 12,960, and Type 3 algae has multiplied by 5,000. So from slowest to fastest, we have Type 3, Type 2, and Type 1.

Module 10: Generating Equivalent Algebraic Expressions *Activity*

A small high school on Hawai'i Island is raising money to buy and plant koa and 'iliahi trees.

Koa sapling

'Iliahi sapling

Below, on the left side are several **classes and a description** of how much money they have each raised. Below, on the right side are several mathematical expressions that describe the **amount of money raised**. The cost of a koa tree is k and the cost of an 'iliahi tree is h.

1. Match each class with at least one expression. A few expressions are equal so **some classes will be matched with more than one expression**.

Freshman class (grade 9):
Each student raised enough money to buy 3 koa and 2 'iliahi trees. There were 7 students.

Sophomore class (grade 10):
The students raised enough money to buy 4 koa and 6 'iliahi trees. Then a local company decided to doubled the money they raised.

Junior class (grade 11):
The students raised enough money to buy 21 koa trees, 14 'iliahi trees, and had $\$21$ left over.

Senior class (grade 12):
The students raised enough money to buy 21 'iliahi trees and had $\$14$ left over.

- $21k + 14h$
- $2(4k + 6h)$
- $7(3k + 2h)$
- $7(3k + 2h + 3)$
- $21k + 14h + 21$
- $12h + 8k$
- $21h + 14$
- $7(2 + 3h)$

2. If each koa tree cost $\$90$, and each ʻiliahi tree cost $\$110$, find the total amount of money that each class raised.

For each class, we have more than one expression that we can use. We may use whichever one is easiest, and we do not need to use them all.

(a) Freshman class

$$\begin{aligned} 21k + 14h &= 21(90) + 14(110) \\ &= 1890 + 1540 \\ &= 3430 \end{aligned}$$

(b) Sophomore class

$$\begin{aligned} 12h + 8k &= 12(110) + 8(90) \\ &= 1320 + 720 \\ &= 2040 \end{aligned}$$

(c) Junior class

$$\begin{aligned} 21k + 14h + 21 &= 21(90) + 14(110) + 21 \\ &= 1890 + 1540 + 21 \\ &= 3451 \end{aligned}$$

(d) Senior class

$$\begin{aligned} 21h + 14 &= 21(110) + 14 \\ &= 2310 + 14 \\ &= 2324 \end{aligned}$$

Unit 4: *Cumulative Activity*

Many animals depend on plants, not just for food, but for shelter as well. For example, the Kamehameha butterfly is an animal that, as a caterpillar, lives on the māmaki plants of Hawaiʻi.

Kamehameha butterfly

Māmaki plant

There are signs that the Kamehameha butterfly is disappearing from many of its natural habitats. Let's grow a field of māmaki plants to help the butterfly thrive.

1. Right now we are building rows of māmaki plants in a 40 feet wide garden. The length of the garden depends on the number of rows. We need **eight feet for every row of māmaki**. For example, if our garden has 4 rows, then it needs a length of 32 feet.

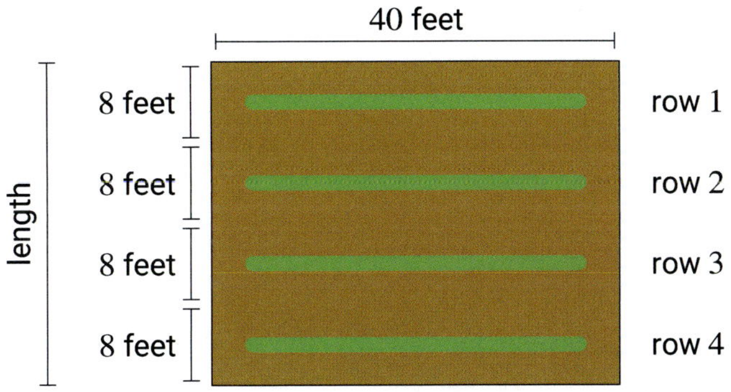

If you are building n rows, write an algebraic expression to show what the length of your garden must be.

Each row has a length of 8 feet so n rows has a length of $8n$ feet.

2. Use the algebraic expression for length (in part 1) to write an expression for the area of the māmaki garden. Check with a friend to see if you have similar answers.

The garden has a length of $8n$ feet and a width of 40 feet. The area of a rectangle is width × length. This gives us an area of $320n$ square feet.

3. Use your expression in part 2 to determine whether the following statements are true or false.

(a) The area is the **sum** of 320 and n. True or **False**

(b) The area is the **product** of 40 and $8n$. **True** or False

(c) The area is the **quotient** of 320 and n. True or **False**

(d) 320 and n are **factors** in your expression in part 2. **True** or False

(e) 40 and $8n$ are **terms** in your expression in part 2. True or **False**

(f) n is a **coefficient** in your expression in part 2. True or **False**

4. You decide to plant 10 rows of māmaki in this garden. What is the total area of the garden? Check with a neighbor to see if you got the same answer.

$$320n = 320 \times 10 = 3200$$

A garden with 10 rows has an area of 3200 square feet.

5. Find the prime factorization of the answer in part 4.

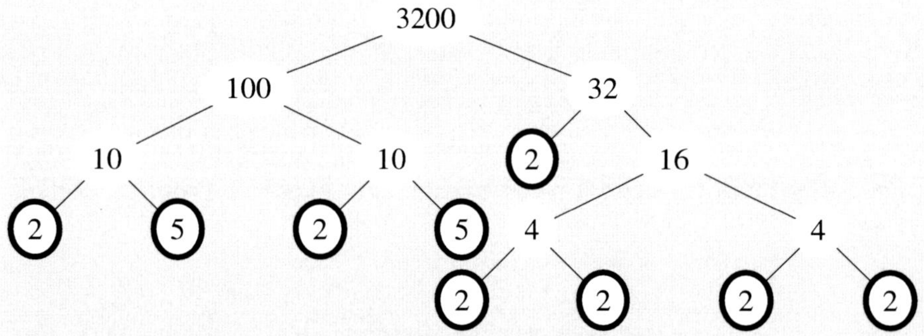

$3200 = 2 \times 2 \times 2 \times 2 \times 2 \times 2 \times 2 \times 5 \times 5 = 2^7 \times 5^2$

6. Work with a partner to sketch, and label the widths and lengths, of at least 3 other rectangles that have the same area as the garden in part 4.

There are twelve possible rectangles with integer sides. The students may also answer with non-integer sides as long as the area is still 3200 square feet.

Possible integer rectangles:
1 feet × 3200 feet
2 feet × 1600 feet
4 feet × 800 feet
5 feet × 640 feet
8 feet × 400 feet
10 feet × 320 feet
16 feet × 200 feet
20 feet × 160 feet
25 feet × 128 feet
32 feet × 100 feet
40 feet × 80 feet
50 feet × 64 feet

Examples:

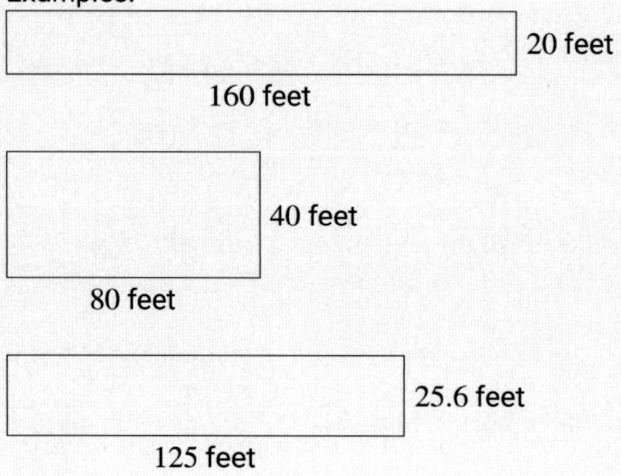

7. How can the prime factorization of part 5 help you in part 6?

To find the sides of the rectangles, we need to factor 3200. Having the prime factorization helps because all factors are made up of the prime factors.

8. Suppose that d number of Kamehameha butterflies land in the garden in part 4, and the Kamehameha butterflies spread out evenly in the area of the garden. Write an expression for the amount of area each of the d Kamehameha butterflies will have to themselves.

The 3200 square feet will be shared among d butterflies. So there will be $\frac{3200}{d}$ square feet per butterfly.

9. How much area would each Kamehameha butterfly have if there were...

(a) 5 Kamehameha butterflies ($d = 5$)? (Round to the nearest tenth if needed.)

The 5 butterflies will have 640 square feet of garden each.

$$\frac{3200}{5} = 640$$

(b) 9 Kamehameha butterflies? (Round to the nearest tenth if needed.)

The 9 butterflies will have 355.555... square feet of garden each. This rounds to 355.6 square feet.

$$\frac{3200}{9} = 355.555... \approx 355.6$$

10. There are 5 Kamehameha butterflies right now. Suppose that the butterfly population doubles every year. How many Kamehameha butterflies would there be in 4 years?

After doubling for 4 years, the population would have multiplied by 2^4. This gives us a population of 80 butterflies.

$$2^4 = 2 \times 2 \times 2 \times 2 = 16$$
$$5 \times 16 = 80$$

Unit 5: Equations and Inequalities

In this unit, we'll learn how to use algebraic equations and inequalities to solve problems through exploring different ahupua'a and the lengths of various Hawaiian trees in the ahupua'a. There are three activities in this unit. *Module 11* involves the use of equations and relationships to help describe the growth of native Hawaiian trees. *Module 12* focuses evaluating the distance between neighbors in a ahupua'a with the help of relationships on two variables. The final activity is cumulative and incorporates concepts from each of the previous activities in this unit.

Common Core State Standards

Common Core State Standard	Module 11	Module 12	Unit 5
6.NS.6 Understand a rational number as a point on the number line. Extend number line diagrams and coordinate axes familiar from previous grades to represent points on the line and in the plane with negative number coordinates.		X	X
6.NS.6b Understand signs of numbers in ordered pairs as indicating locations in quadrants of the coordinate plane; recognize that when two ordered pairs differ only by signs, the locations of the points are related by reflections across one or both axes.		X	
6.NS.6c Find and position integers and other rational numbers on a horizontal or vertical number line diagram; find and position pairs of integers and other rational numbers on a coordinate plane.		X	X
6.NS.8 Solve real-world and mathematical problems by graphing points in all four quadrants of the coordinate plane. Include use of coordinates and absolute value to find distances between points with the same first coordinate or the same second coordinate.		X	
6.EE.5 Understand solving an equation or inequality as a process of answering a question: which values from a specified set, if any, make the equation or inequality true? Use substitution to determine whether a given number in a specified set makes an equation or inequality true.	X		X
6.EE.6 Use variables to represent numbers and write expressions when solving a real-world or mathematical problem; understand that a variable can represent an unknown number, or, depending on the purpose at hand, any number in a specified set.	X		X
6.EE.7 Solve real-world and mathematical problems by writing and solving equations of the form $x + p = q$ and $px = q$ for cases in which p, q and x are all nonnegative rational numbers.		X	X

Continued on next page

Common Core State Standard	Module 11	Module 12	Unit 5
6.EE.8 Write an inequality of the form $x > c$ or $x < c$ to represent a constraint or condition in a real-world or mathematical problem. Recognize that inequalities of the form $x > c$ or $x < c$ have infinitely many solutions; represent solutions of such inequalities on number line diagrams.	X		
6.EE.9 Use variables to represent two quantities in a real-world problem that change in relationship to one another; write an equation to express one quantity, thought of as the dependent variable, in terms of the other quantity, thought of as the independent variable. Analyze the relationship between the dependent and independent variables using graphs and tables, and relate these to the equation. *For example, in a problem involving motion at constant speed, list and graph ordered pairs of distances and times, and write the equation $d = 65t$ to represent the relationship between distance and time.*	X	X	X

Module 11: Equations and Relationships *Activity*

This house is built right next to a forest. Near the house are some fast-growing trees.

After x years, the height of the trees (in feet) are given by the following expressions.

Tree	Height after x years
Mai'a (banana)	$2x+2$
Loulu (palm)	$\frac{3}{2}x$
Niu (coconut)	$3x+1$
'Ohe (bamboo)	$\frac{5}{2}x+3$

1. When writing equations and inequalities about these trees, we can replace "the height of the tree" with the expression that describes it.
   For example,
   "the height of the 'ohe is greater than or equal to 15 feet" can be written as
   $$\frac{5}{2}x+3 \geq 15$$
   and
   "the height of the 'ohe is the height of the niu" can be written as
   $$\frac{5}{2}x+3 = 3x+1.$$
   Write the equations and inequalities for each of the following sentences. You **do not** need to solve them.

   (a) The height of the loulu is 3 feet.

   $\frac{3}{2}x = 3$

   (b) The height of the niu is half of the height of the mai'a.

   $3x+1=\frac{1}{2}(2x+2)$ or $3x+1=x+1$

   (c) The height of the mai'a is less than the height of the 'ohe.

   $2x+2<\frac{5}{2}x+3$

   (d) The height of the niu is 2 feet more than the height of the 'ohe.

   $3x+1=\left(\frac{5}{2}x+3\right)+2$ or $3x+1=\frac{5}{2}x+5$

2. How tall will each tree be after 2, 4, and 6 years?

	Tree height (feet)			
Number of years	Mai'a	Loulu	Niu	'Ohe
x	$2x+2$	$\frac{3}{2}x$	$3x+1$	$\frac{5}{2}x+3$
2	$2\times(2)+2$ $=6$	3	$3\times(2)+1$ $=7$	8
4	$2\times(4)+2$ $=10$	$\frac{3}{2}\times(4)$ $=6$	13	$\frac{5}{2}\times(4)+3$ $=13$
6	14	$\frac{3}{2}\times(6)$ $=9$	$3\times(6)+1$ $=19$	$\frac{5}{2}\times(6)+3$ $=18$

3. Draw the trees after 6 years ($x=6$).

4. There is a 15-foot tall power line connected to this house. We need to make sure that the trees doesn't grow as tall as the power line.

   (a) Which tree(s) would reach the power line in 6 years?

   We need to worry about the niu and the 'ohe because their height will be over 15 feet in 6 years.

   (b) What are three things that we can do to the trees if they get too tall?

Module 12: Relationships in Two Variables *Activity*

You and your ʻohana (family) live in this ahupuaʻa. Here is a map of your ahupuaʻa.

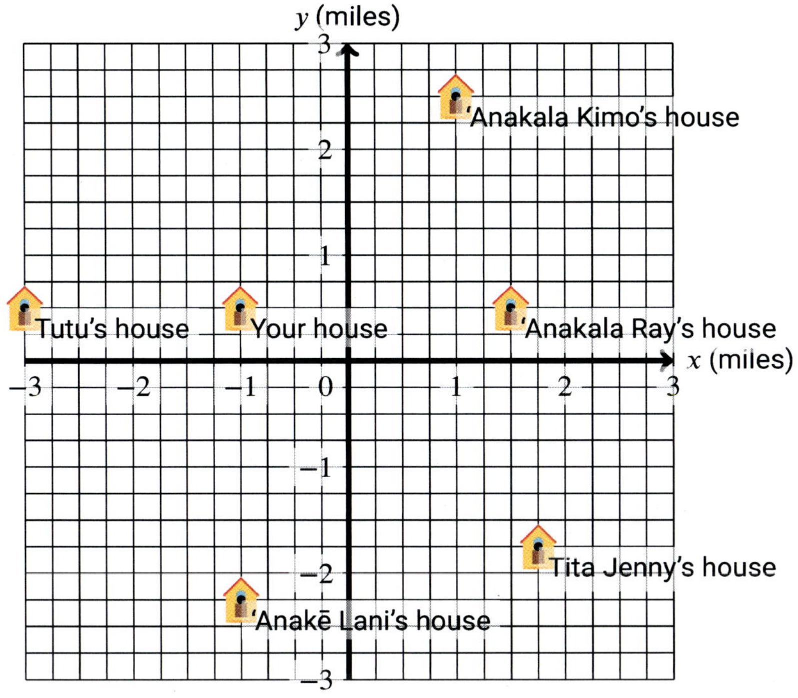

1. If you only move left, right, up, or down (not diagonal or in a curve), how many miles away **from your house** does each relative live?

   To answer this, we first look at the position of your house $(-1, 0.5)$ and a relative's house. Next, we find the difference between the x coordinates and difference between the y coordinates. Finally, we add the absolute values of these differences.

   (a) Tutu

   This house is at $(-3, 0.5)$ which is 2 miles from your house.

   $|-3-(-1)|+|0.5-0.5| = 2+0 = 2$

   (b) ʻAnakala Ray

   This house is at $(1.5, 0.5)$ which is 2.5 miles from your house.

   $|1.5-(-1)|+|0.5-0.5| = 2.5+0 = 2.5$

   (c) ʻAnakē Lani

   This house is at $(-1, -2.25)$ which is 2.75 miles from your house.

   $|-1-(-1)|+|-2.25-0.5| = 0+2.75 = 2.75$

   (d) ʻAnakala Kimo

   This house is at $(1, 2.5)$ which is 4 miles from your house.

   $|1-(-1)|+|2.5-0.5| = 2+2 = 4$

   (e) Tita Jenny

   This house is at $(1.75, -1.75)$ which is 5 miles from your house.

   $|1.75-(-1)|+|-1.75-0.5| = 2.75+2.25 = 5$

2. Suppose that you can jog 4 miles per hour. Write an equation describing the distance you jog (d) after t hours.

$d = 4t$

3. Using your equation from part 2 and the distances from part 1, calculate how many hours, t, it takes to jog from your house to...

   (a) Tutu's house

$$\begin{array}{rcl} 2 & = & 4t \\ \div 4 & & \div 4 \\ 0.5 & = & t \end{array}$$

It will take 0.5 hours to jog to Tutu's house.

   (b) ʻAnakala Ray's house

$$\begin{array}{rcl} 2.5 & = & 4t \\ \div 4 & & \div 4 \\ 0.625 & = & t \end{array}$$

It will take 0.625 hours to jog to ʻAnakala Ray's house.

   (c) ʻAnakē Lani's house

$$\begin{array}{rcl} 2.75 & = & 4t \\ \div 4 & & \div 4 \\ 0.6875 & = & t \end{array}$$

It will take 0.6875 hours to jog to ʻAnakē Lani's house.

   (d) ʻAnakala Kimo's house

$$\begin{array}{rcl} 4 & = & 4t \\ \div 4 & & \div 4 \\ 1 & = & t \end{array}$$

It will take 1 hour to jog to ʻAnakala Kimo's house.

   (e) Tita Jenny's house

$$\begin{array}{rcl} 5 & = & 4t \\ \div 4 & & \div 4 \\ 1.25 & = & t \end{array}$$

It will take 1.25 hours to jog to Tita Jenny's house.

Unit 5: *Cumulative Activity*

A pond in your ahupua'a is badly polluted. You want to clean it up but, it is big pond and you wont be able to clean it alone so you decide to ask some of your neighbors to help out.

1. If x is the number of neighbors who will help, write an expression for the total number of workers (including yourself) who will clean up the pond.

   $x+1$

2. The percent of pollution remaining in the pond can be given by a simple algebraic equation: "The *percent of pollution* is 100% divided by the *total number of workers* cleaning it up." Write this algebraic equation for percent of pollution, y in terms of number of neighbors, x.

   $$y = \frac{100}{x+1}$$

3. Complete the following table (round to one decimal place, if needed).

Number of neighbors, x	Percent polluted, y
0	100
1	50
2	33.3
3	25
4	20
5	16.7
6	14.3
7	12.5
8	11.1
9	10
10	9.1

4. Graph the data from the previous table.

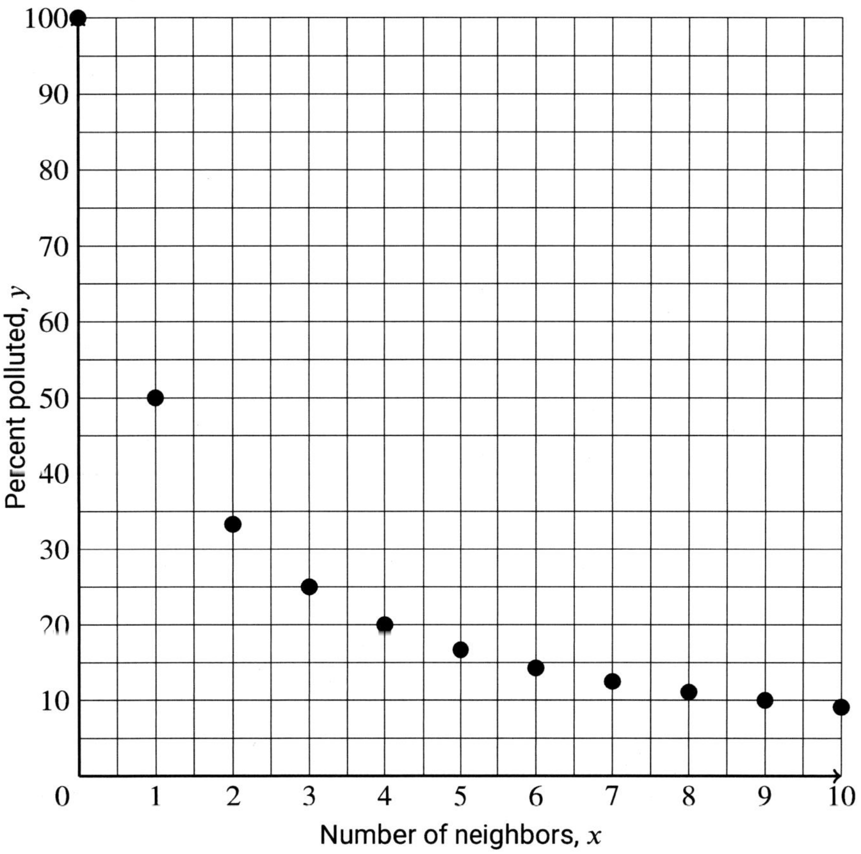

5. The graph in part 4 only shows Quadrant I. Do you think there is important information in any of the other three quadrants? Why or why not? Feel free to discuss with your friends.

No. In Quadrant II and III, x is negative. This does not make sense because we will not have a negative number of neighbors helping us. In Quadrant IV, y is negative. This does not make sense either, because we will not be able to bring pollution down to a negative number. At best, it would be 0% polluted.

Due to the pollution, there hasn't been a lot of animals visiting this pond.

6. Birds will return to the pond if it is less than 15% polluted. Modify the equation from part 2 to describe this as an inequality in terms of x.

$$\frac{100}{x+1} < 15$$

7. Use the previous table or graph to answer this question. How many neighbors need to help in order to get this pond clean enough for birds to return?

The pollution is over 15% when $x = 5$, but under 15% when $x \geq 6$. So we need at least 6 neighbors cleaning.

At the end of the cleanup effort, the birds started to return to the pond. Here are the number of birds, b, d days after the cleanup.

Days (d)	0	1	2	3	4	5	6
Birds (b)	0	3	6	9	12	15	18

8. Which variable, b or d, is the independent variable, and which variable is the dependent variable? How do you know?

d is the independent variable and b is the dependent variable. We know this because we are trying to measure b and how it changes as s changes.

9. Write the equation for b in terms of d.

$b = 3d$

10. With a partner or in the online comment section, describe a place that you have been to that was really polluted. Describe what it would look like if people took the time to clean it up.

Unit 6: Relationships in Geometry

In this unit, we'll learn how to draw shapes on a grid and calculate length, area, and surface area through surveying Hawaiian land, exploring the lengthy process of kapa making, and building a hale. There are four activities in this unit. *Module 13* involves helping a land surveyor with the use of area and polygons. *Module 14* focuses on distance and area in the coordinate plane to help wahine make kapa. *Module 15* explores how to build a hale with the help of surface area and volume of solids. The final activity is cumulative and incorporates concepts from each of the previous activities in this unit.

For some of the activities in this unit, a yard stick or measuring tape would be helpful but is not required.

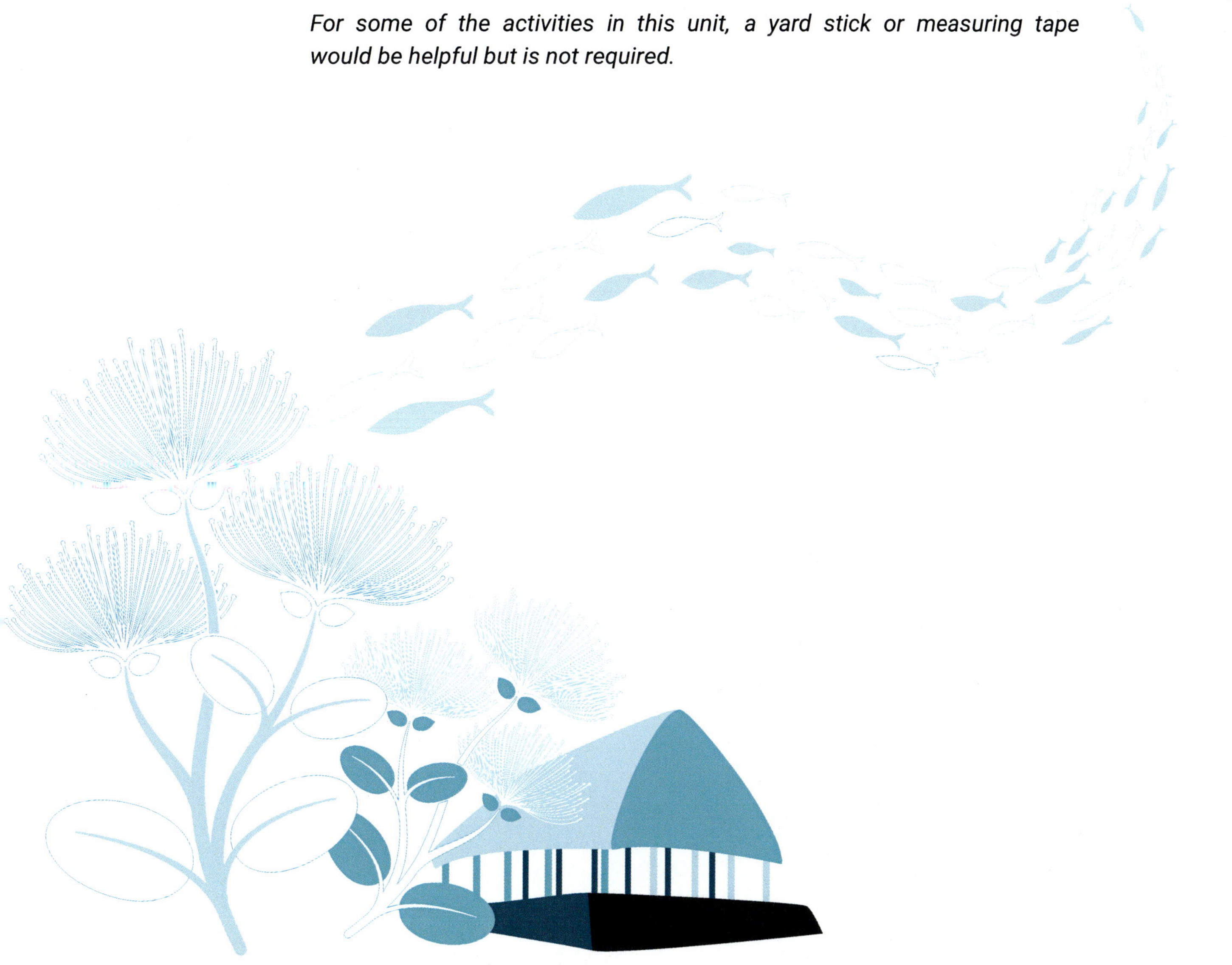

Common Core State Standards

Common Core State Standard	Module 13	Module 14	Module 15	Unit 6
6.NS.6b Understand signs of numbers in ordered pairs as indicating locations in quadrants of the coordinate plane; recognize that when two ordered pairs differ only by signs, the locations of the points are related by reflections across one or both axes.		X		X
6.NS.8 Solve real-world and mathematical problems by graphing points in all four quadrants of the coordinate plane. Include use of coordinates and absolute value to find distances between points with the same first coordinate or the same second coordinate.		X		
6.EE.7 Solve real-world and mathematical problems by writing and solving equations of the form x + p = q and px = q for cases in which p, q and x are all nonnegative rational numbers.	X			
6.G.1 Find area of right triangles, other triangles, special quadrilaterals, and polygons by composing into rectangles or decomposing into triangles and other shapes; apply these techniques in the context of solving real-world and mathematical problems.	X	X	X	X
6.G.2 Find the volume of a right rectangular prism with fractional edge lengths by packing it with unit cubes of the appropriate unit fraction edge lengths, and show that the volume is the same as would be found by multiplying the edge lengths of the prism. Apply the formulas V = l w h and V = b h to find volumes of right rectangular prisms with fractional edge lengths in the context of solving real-world and mathematical problems.			X	
6.G.3 Draw polygons in the coordinate plane given coordinates for the vertices; use coordinates to find the length of a side joining points with the same first coordinate or the same second coordinate. Apply these techniques in the context of solving real-world and mathematical problems.		X		X

Continued on next page

Common Core State Standard	Module 13	Module 14	Module 15	Unit 6
6.G.4 Represent three-dimensional figures using nets made up of rectangles and triangles, and use the nets to find the surface area of these figures. Apply these techniques in the context of solving real-world and mathematical problems.			X	X

Module 13: Area and Polygons *Activity*

A land surveyor is a person who measures and maps out the details of a piece of land. Let's survey two pieces of land.

The first piece of land is from the ahupua'a of Wai'anae. We made it easier to measure the land by breaking it up into right triangles and rectangles. Some of the sides and areas (the numbers in a circle) have been measured.

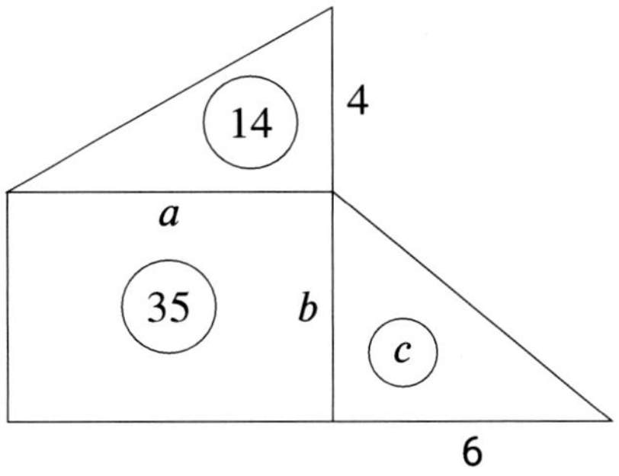

1. Find the two missing sides and then the missing area.

   (a) Side a:

   At the top, is a right triangle with a base of a, height of 4, and area of 14.

   $$\begin{aligned} \tfrac{1}{2}\times a\times 4 &= 14 \\ 2\times a &= 14 \\ \div 2 & \quad \div 2 \\ a &= 7 \end{aligned}$$

   Using the formula for the area of triangle, we see that $a = 7$.

   (b) Side b:

   We have a rectangle with a width of a, a height of b, and an area of 35. We previously found that $a = 7$.

   $$\begin{aligned} a\times b &= 35 \\ 7\times b &= 35 \\ \div 7 & \quad \div 7 \\ b &= 5 \end{aligned}$$

   Using the formula for the area of rectangle, we see that $a = 5$.

   (c) Area c:

   To the right is a right triangle with a base of 6, height of b, and area of c. We previously found that $b = 5$.

   $$\begin{aligned} \tfrac{1}{2}\times 6\times b &= c \\ \tfrac{1}{2}\times 6\times 5 &= c \\ 15 &= c \end{aligned}$$

   Using the formula for the area of triangle, we see that $c = 15$.

2. Find the total area of the land in Wai'anae.

   The three areas are 14, 35 and 15. This totals 64.

Here is a piece of land from the ahupua'a of Waiau.

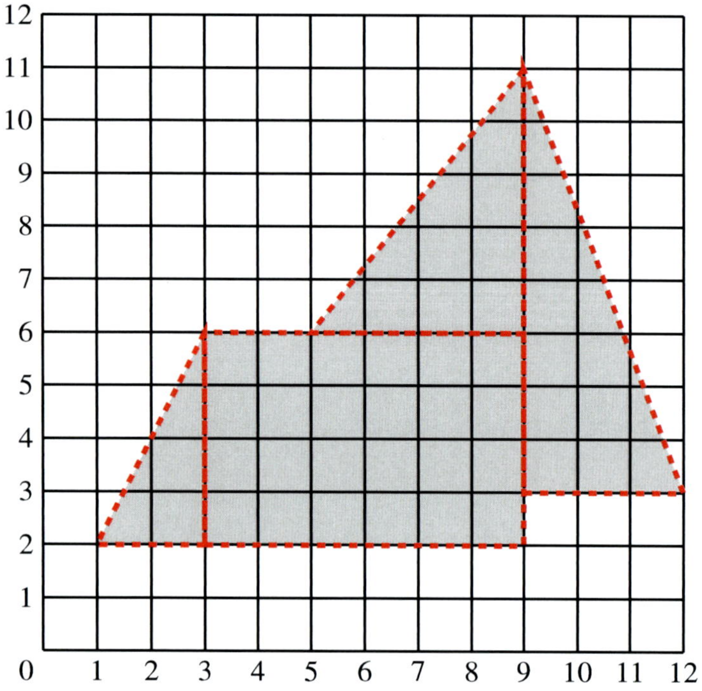

3. Break up the Waiau land into right triangles and rectangles by drawing on the figure above.

   There are many different ways to do this. Above is just one example.

4. Find the total area of the land in Waiau.

   The figure is made up of four shapes. A rectangle in the middle surrounded by three triangles: on the left, on the top, and on the right.

   The rectangle has a width of 6 and a height of 4.
   Area of rectangle $= 6 \times 4 = 24$

   The left triangle has a width of 2 and a height of 4. The top triangle is 4 by 5. The triangle on the right is 3 by 8.
   Area of triangle on left $= \frac{1}{2} \times 2 \times 4 = 4$
   Area of triangle on top $= \frac{1}{2} \times 4 \times 5 = 10$
   Area of triangle on right $= \frac{1}{2} \times 3 \times 8 = 12$

   Total area $= 24 + 4 + 10 + 12 = 50$

   The total area is 50.

5. Which piece has a larger total area, Wai'anae or Waiau?

   The piece from Wai'anae has a larger total area.

Module 14: Distance and Area in the Coordinate Plane *Activity*

Kapa is a cloth made from the wauke (paper mulberry) plant. It takes a lot of hard work to make kapa. In old Hawai'i, Hawaiians gathered strips of young wauke bark and repeatedly battered it until the fabric was soft. Then they had to forage for more pieces, line them up next to each other, and pound the edges until the different strips connected. It would take a very long time to make a large piece that was soft, strong, flat, and had the same thickness throughout. After the pieces were beaten together, they had to be dried in the sun before being made into clothing, furniture, or wraps.

The making of kapa was very important to the ancient Hawaiians. It was said that the demi-god, Maui, threw a hook into the Sun to slow it down so that his mom, Hina, could dry her kapa that she had worked so hard to make.

A wahine in our village is making kapa and she needs a place to dry it in the sun. However, we need to build a fence to keep animals and kolohe kids away from the kapa while it is drying.

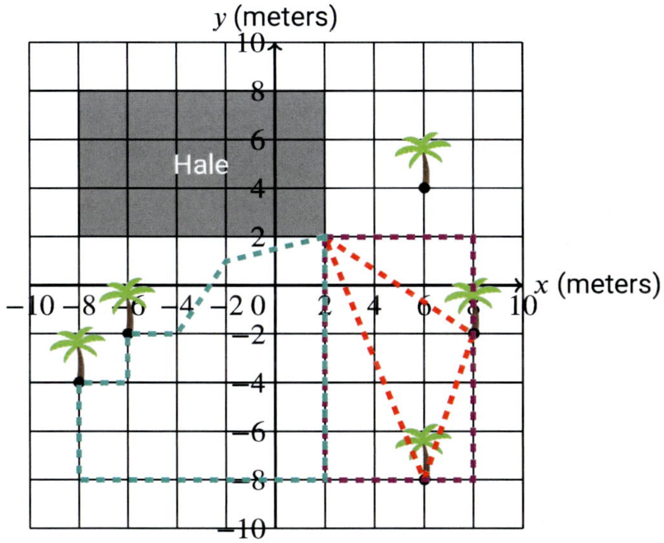

1. Draw a triangle connecting the three points on the map above: $(2, 2)$, $(8, -2)$, and $(6, -8)$.
2. Let's draw a rectangle around the triangle in part 1 by connecting the following points: $(2, 2)$, $(8, 2)$, $(8, -8)$, and $(2, -8)$. You should now see four triangles: the original one and three right triangles.

   The triangle is done in red and the rectangle is done in purple above.

3. Use the rectangle and the three right triangles in part 2 to find the area of the original triangle in part 1. Show your work below.

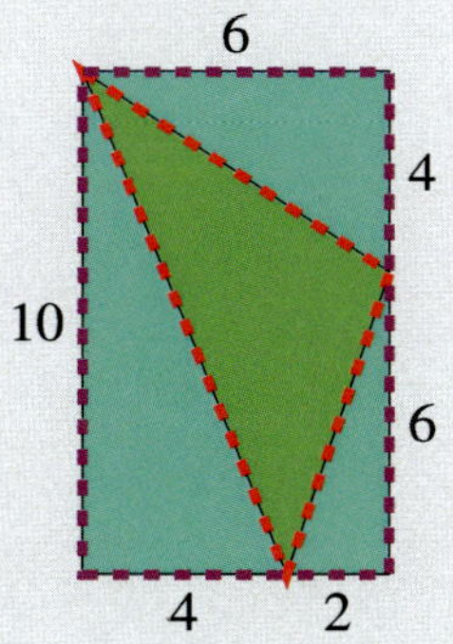

If we take the area of the rectangle and subtracted the total area of the three right triangles, then we can find the area of the original triangle.

Rectangle:

$6 \times 10 = 60$

Right triangles:

$\frac{1}{2} \times 4 \times 10 = 20$

$\frac{1}{2} \times 6 \times 4 = 12$

$\frac{1}{2} \times 2 \times 6 = 6$

Area of the triangle $= 60 - (20 + 12 + 6) = 60 - 38 = 22$

The area of the triangle is 22 square meters.

4. The wahine needs at least 40 square meters to dry all of the kapa. Is there enough area in the triangle?

No, the triangle is only 22 square meters.

5. Draw a new drying area by connecting the following points in order.
$(2,2) \rightarrow (-2,1) \rightarrow (-4,-2) \rightarrow (-6,-2) \rightarrow (-6,-4) \rightarrow (-8,-4) \rightarrow (-8,-8) \rightarrow (2,-8) \rightarrow (2,2)$

This is done in blue on the original grid with the hale.

6. Find the area of the polygon that you drew in part 5.

There are many ways to break this polygon up into smaller shapes. Below is an example where the polygon is broken into two triangles and two rectangles.

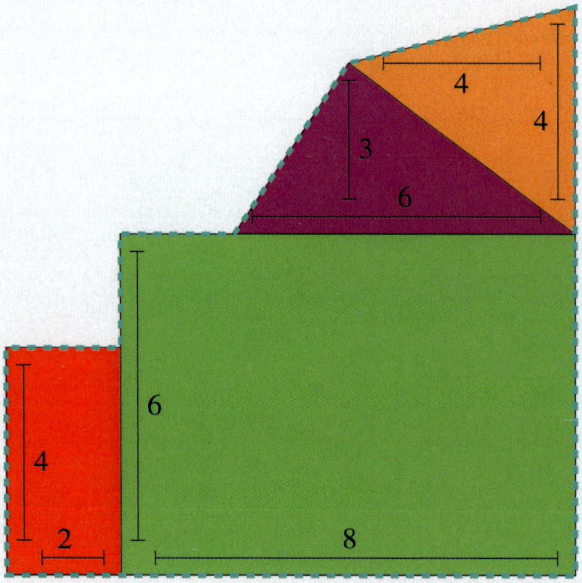

Areas of triangles:
$\frac{1}{2} \times 4 \times 4 = 8$
$\frac{1}{2} \times 6 \times 3 = 9$

Areas of rectangles:
$2 \times 4 = 8$
$8 \times 6 = 48$

Total area:
$8 + 9 + 8 + 48 = 73$

The total area of this polygon is 73 square meters.

7. Is there enough area in this polygon for the wahine to dry her kapa?

Yes, the polygon has more than the 40 square meters needed.

Module 15: Surface Area and Volume of Solids *Activity*

A yard stick or measuring tape could be helpful in this activity.

In parts of Hawai'i that stay warm all year long, the ancient Hawaiians rarely spent much time in their hale. The doors on hale were often made very small to keep pets and pests out, and the insides of the hale were often very warm and dark. The hale was mostly used for storage and rituals, except for cold and rainy weather or special occasions.

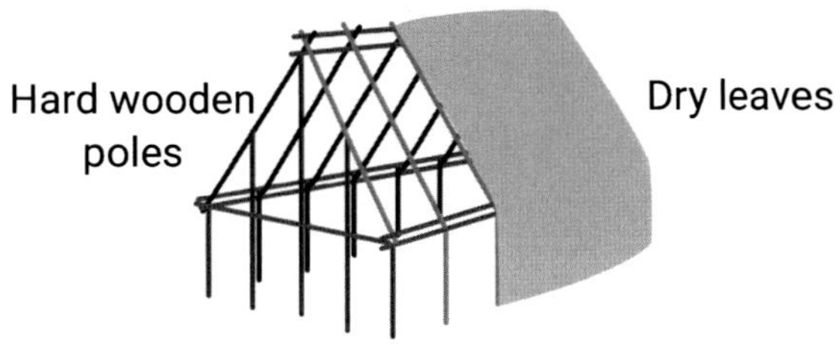

A hale under construction

We are building a hale. Let's first look at it without a roof.

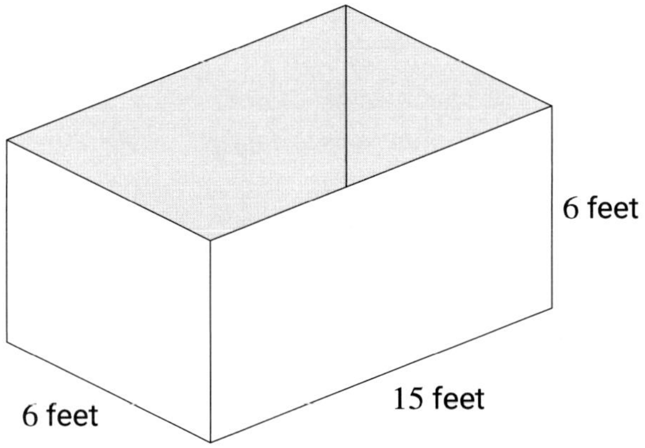

1. Find the volume of the rectangular prism.

   $(6 \text{ feet}) \times (15 \text{ feet}) \times (6 \text{ feet}) = 540 \text{ feet}^3$

   Without the roof, this hale has a volume of 540 cubic feet.

2. How many of your classmates do you think can fit in a hale of this size? Discuss with a friend. Walking around the classroom with a yard stick or measuring tape while imagining a space this size can also help.

3. Divide your answer in part 1 by your answer in part 2 to find out how much space each of your classmates would take up in this hale. Round to the nearest tenth.

Now let's look at the hale with its roof.

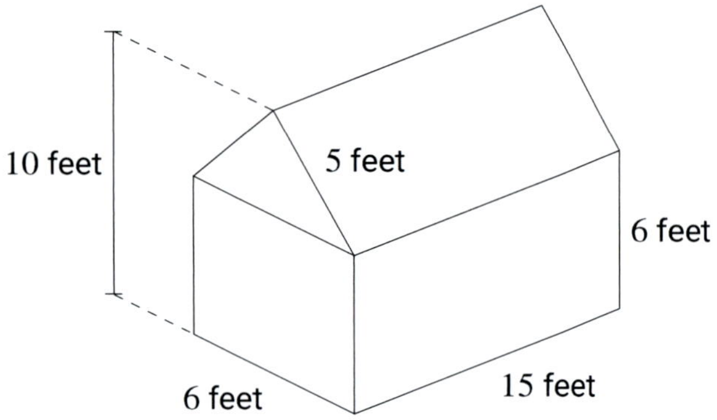

4. Sketch the four walls and the four sides of the roof as a geometric net. Label each edge with its length. You only need to use triangles, squares, and rectangles.

There are many ways to draw the geometric. Here is just one.

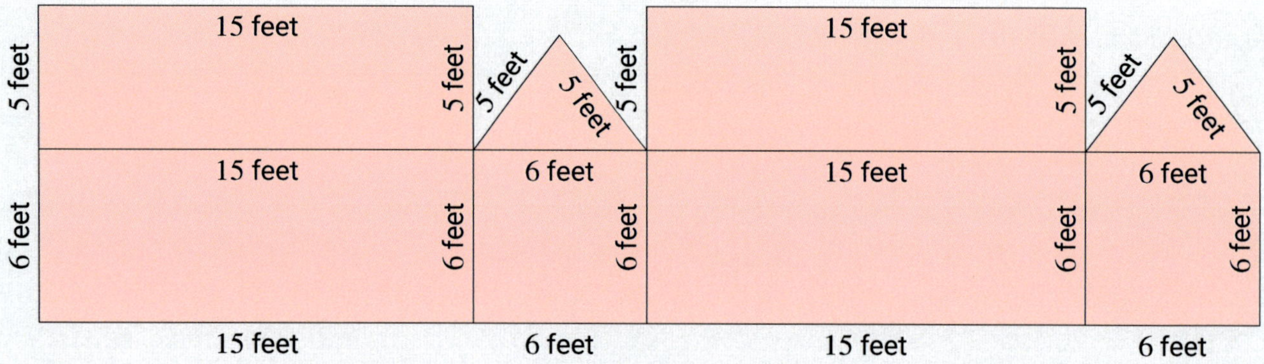

5. We need to cover the four walls and the four sides of the roof with thick leaves. Find the total surface area of these eight pieces.

There are four rectangles, two squares, and two triangles.

Area of the rectangle:
$6 \times 15 = 90$

Area of the square:
$6 \times 6 = 36$

Area of the triangle:
The 3D image of the hale shows us that the height of the triangles is 4 feet.

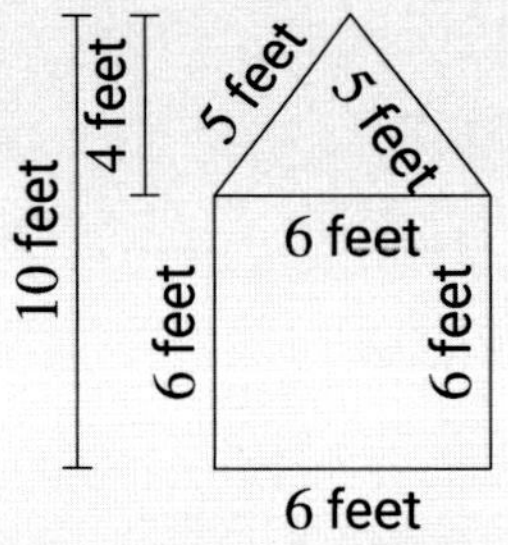

$\frac{1}{2} \times 6 \times 4 = 12$

Total area:
$4 \times 90 + 2 \times 36 + 2 \times 12 = 360 + 72 + 24 = 456$

The total surface area is 456 square feet.

6. When a hale needs to be built, many people must come together to build it. When was the last time you built something with your hands? What did you make?

Unit 6: *Cumulative Activity*

Ivy gourd is an invasive vine that covers and kills other plants. A bird pooped in a field next to your ʻili (neighborhood) and three ivy gourd plants are starting to grow from the seeds in the bird poop.

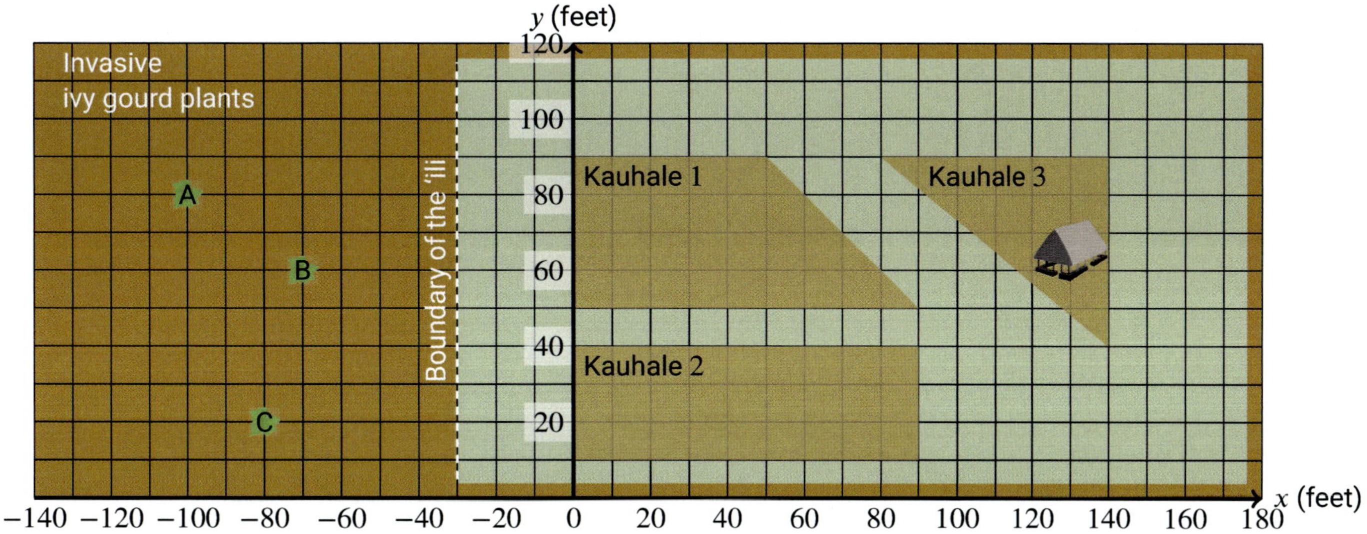

1. How far away is each plant from the boundary of the ʻili?

   The boundary of the ʻili is at $x = -30$ (feet). To find the distances from the boundary, we look at its x value to see how much it differs from -30.

   (a) Plant A: $|-100-(-30)| = |-(100-30)| = |-70| = 70$ Plant A is 70 feet away from the boundary.

   (b) Plant B: $|-70-(-30)| = |-(70-30)| = |-40| = 40$ Plant B is 40 feet away from the boundary.

   (c) Plant C: $|-80-(-30)| = |-(80-30)| = |-50| = 50$ Plant C is 50 feet away from the boundary.

2. Look at the coordinates of each ivy gourd plant. If we changed the signs of each x-coordinate to it's opposite, would the plant be moved into a kauhale? If so, which kauhale would it be in?

   (a) Plant A: Yes. Plant A's coordinates would be reflected to $(100, 80)$ which is in kauhale 3.

   (b) Plant B: Yes. Plant B's coordinates would be reflected to $(70, 60)$ which is in kauhale 1.

   (c) Plant C: Yes. Plant C's coordinates would be reflected to $(80, 20)$ which is in kauhale 2.

3. After the ivy gourd plants enter your 'ili, they quickly cover the three kauhale, where the hale (houses) are built. What is the **total** area of the three kauhale?

Kauhale 1:

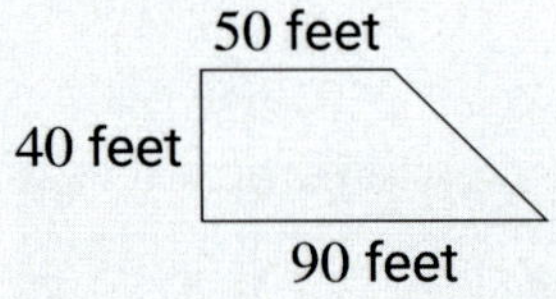

$\frac{1}{2} \times (90 + 50) \times 40 = 2800$

Kauhale 2:

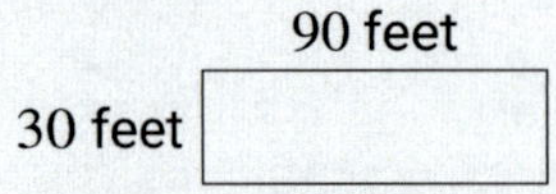

$90 \times 30 = 2700$

Kauhale 3:

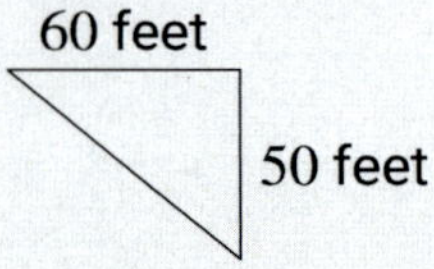

$\frac{1}{2} \times 60 \times 50 = 1500$

Total area $= 2800 + 2700 + 1500 = 7000$

The total area of the three kauhale is $7,000$ square feet.

4. In kauhale 3, the ivy gourd is invading a hale. First, it covers the floor of the hale, then the walls and the roof. What is the surface area of the covered hale?

Hale

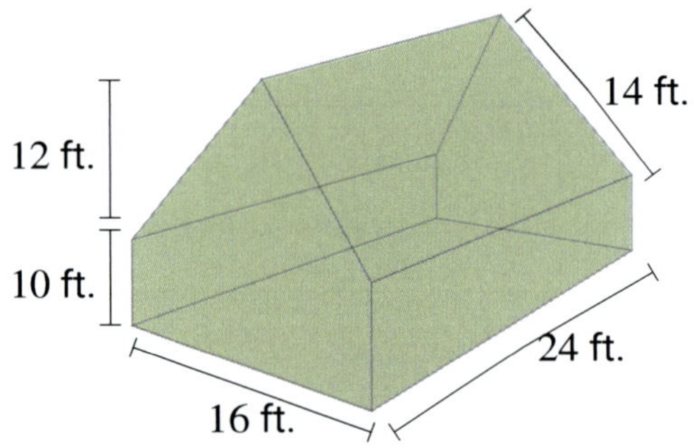

Hint: The front and back walls is a triangle on top of a rectangle. The roofs, side walls, and floor are rectangles.

Floor:

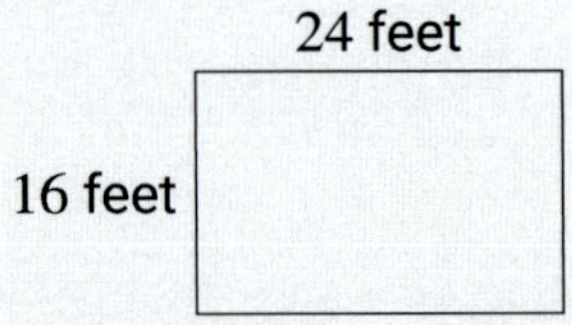

$16 \times 24 = 384$

Side walls (two of these):

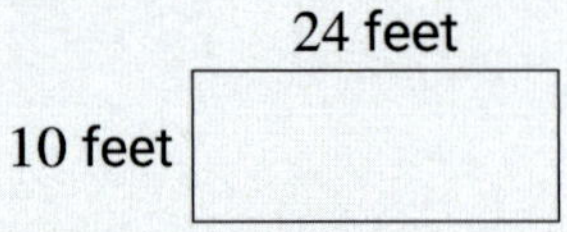

$10 \times 24 = 240$

Roofs (two of these):

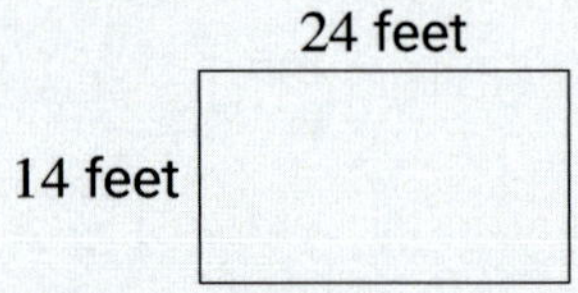

$14 \times 24 = 336$

Front and back pentagonal walls (two of these):

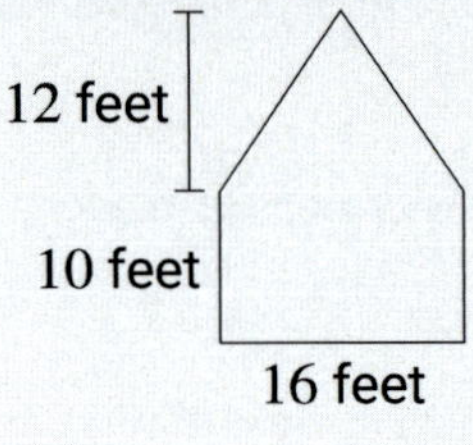

This can be decomposed into a triangle and a rectangle.

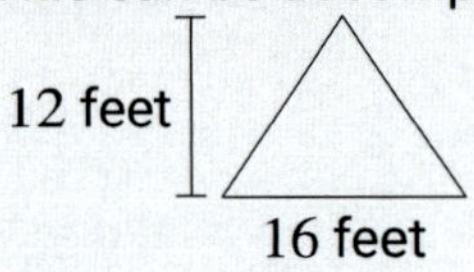

$\frac{1}{2} \times 16 \times 12 = 96$

$10 \times 16 = 160$

$$\begin{aligned}\text{Total area} &= \text{Floor} + 2 \times \text{Side walls} + 2 \times \text{Roofs} + 2 \times \text{Front and back walls} \\ &= (384) + 2 \times (240) + 2 \times (336) + 2 \times (96 + 160) \\ &= (384) + 2 \times (240) + 2 \times (336) + 2 \times (256) \\ &= 2048\end{aligned}$$

The total surface area of the hale is $2,048$ square feet.

5. Most plants are good for the environment and good for people, but certain plants, when grown in wrong places, can cause a lot of trouble. Describe a plant in your neighborhood or school that you think is out of place. Describe why you think it might be bad for your neighborhood. You may also find or draw a picture of it. Please share with your partner or in the online comment section. (If sharing online, please don't post pictures of people's faces or anything that can be used to find a personal address.)

Unit 7: Measurement and Data

In this unit, we'll learn how to use statistics to summarize and display data through evaluating the shape and length of a Hawaiian spear and the exploration of the Big Island. There are two activities in this unit. *Module 16* explores the Hawaiian spear, pololū, through displaying, analyzing, and summarizing data. The final activity is cumulative and incorporates concepts from each of the previous activities in this unit.

Common Core State Standards

Common Core State Standard	Module 16	Unit 7
6.SP.1 Recognize a statistical question as one that anticipates variability in the data related to the question and accounts for it in the answers. *For example, "How old am I?" is not a statistical question, but "How old are the students in my school?" is a statistical question because one anticipates variability in students' ages.*	X	X
6.SP.2 Understand that a set of data collected to answer a statistical question has a distribution which can be described by its center, spread, and overall shape.	X	X
6.SP.3 Recognize that a measure of center for a numerical data set summarizes all of its values using a single number, while a measure of variation describes how its values vary using a single number.	X	X
6.SP.4 Display numerical data in plots on a number line, including dot plots, histograms, and box plots.	X	X
6.SP.5 Summarize numerical data sets in relation to their context, such as by:	X	X
6.SP.5a Reporting the number of observations.	X	X
6.SP.5b Describing the nature of the attribute under investigation, including how it was measured and its units of measurement.	X	X
6.SP.5c Giving quantitative measures of center (median and/or mean) and variability (interquartile range and/or mean absolute deviation), as well as describing any overall pattern and any striking deviations from the overall pattern with reference to the context in which the data were gathered.	X	X
6.SP.5d Relating the choice of measures of center and variability to the shape of the data distribution and the context in which the data were gathered.	X	X

Module 16: Displaying, Analyzing, and Summarizing Data *Activity*

The ancient Hawaiians were very fortunate to have plenty of woody plants around them. The wood from koa trees and kauila shrubs, in particular, are very solid. This quality made them very important for making tools and weapons in ancient Hawai'i. Let's use statistics to figure out the typical length of a Hawaiian spear called the pololū.

Pololū

There is a collection of ancient pololū in a museum. Let's take a look at their lengths (in feet).

11	15	8	7	11	10
9	6	12	10	13	10
9	9	9	8	7	12

1. Plot the lengths on a dot plot.

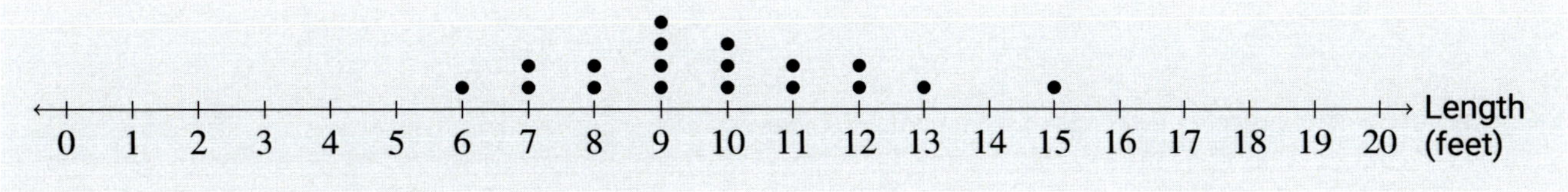

2. Describe the shape of this data distribution.

The data distribution is in the shape of a triangle or bell. The highest peak is at 9 feet. We did not find any pololū that were shorter than 6 feet or longer than 15 feet.

3. Which pololū length was the most common? How many of the ancient pololū were this length?

Pololū length of 9 feet was most common with 4 spears.

4. Find the mean and median of this data. Round to the nearest tenth.

Mean:

$$\frac{6+7+7+8+8+9+9+9+9+10+10+10+11+11+12+12+13+15}{18} = \frac{176}{18} = 9.777...$$

≈ 9.8

Median:

6 7 7 8 8 9 9 9 (9) (10) 10 10 11 11 12 12 13 15

(9.5)

The mean length is 9.8 feet and the median length is 9.5 feet.

5. Could any of your answers from parts 3 and 4 be used to represent the typical length of a pololū? Why or why not? If not, which one of those answers do you think is best?

   The data has a distribution that is mostly symmetrical. There is an outlier at 15 feet, but it is not dramatically different from the rest of the data. As a result, the answers from parts 3 and 4 are very similar and any of the measurements 9, 9.8, or 9.5 feet can be used to represent the typical length of a pololū.

6. How do you think life would have been different in ancient Hawai'i if woody plants were rare? Please share with your partner or in the online comment section.

Unit 7: *Cumulative Activity*

King Kamehameha the Great was born in the moku of Kohala, on the northern end of Hawai'i Island. Today, Kohala is split into North Kohala and South Kohala. Let's look at the sizes of some ahupua'a in North Kohala.

Ahupua'a	Area (km^2)
'A'amakāō	7.06
'Āinakea	2.18
'Āpuakaohau	1.19
'Āwini	8.00
Hala'ula	2.45
Hālawa	6.42
Halelua	3.03
Hana'ula	2.53
Hāwī	1.29
Honokāne	19.03
Honomaka'u	1.46
Honopueo	3.30

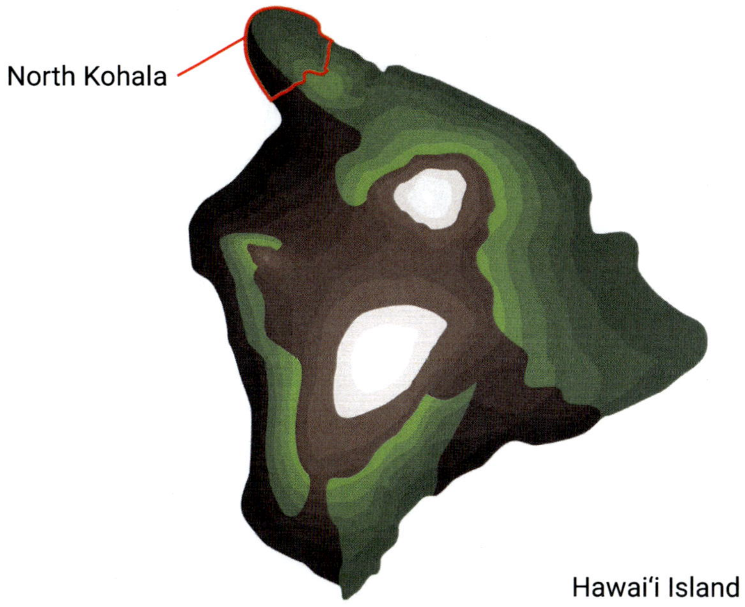

1. Make a box and whisker graph using the area of each ahupua'a.

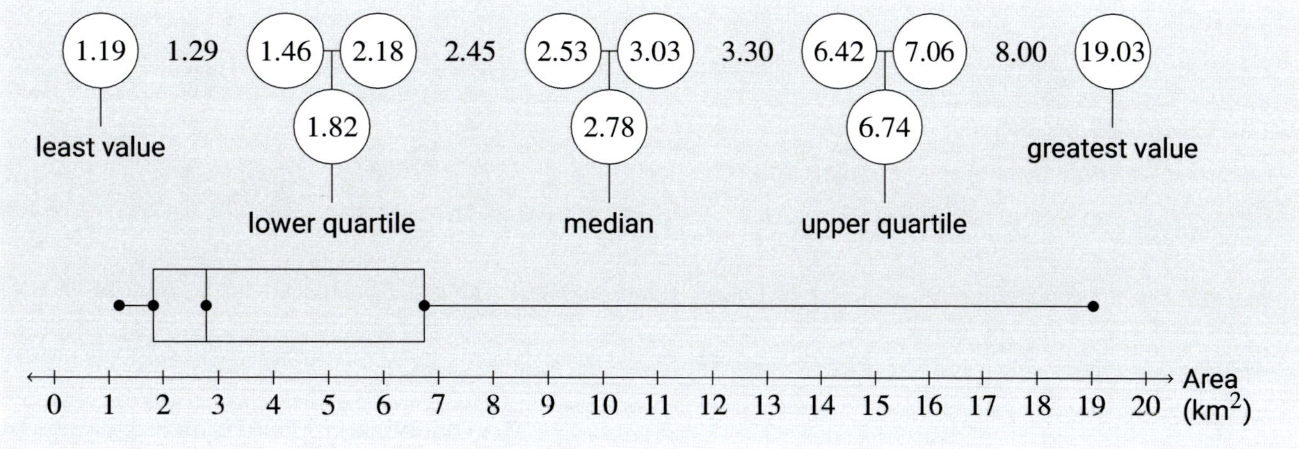

2. Calculate the range and interquartile range of the areas.

Range = greatest value − least value = $19.03 - 1.19 = 17.84$

Interquartile range = upper quartile − lower quartile = $6.74 - 1.82 = 4.92$

3. Calculate the mean of the areas. Round to two decimal places.

$1.19 + 1.29 + 1.46 + 2.18 + 2.45 + 2.53 + 3.03 + 3.30 + 6.42 + 7.06 + 8.00 + 19.03 = 57.94$
$57.94 \div 12 \approx 4.83$

The mean size of the ahupua'a in North Kohala is 4.83 square kilometers.

4. If we removed Honokāne from the table, how would this affect your box and whisker graph?

   Without Honokāne, the greatest value would be much smaller. In fact, the next greatest value is much smaller than the value of Honokāne. So the long line representing the highest quartile would be much shorter if Honokāne was removed.

5. Would removing Honokāne from your table affect the mean? If so, how would it be affected? (You do not need to calculate the mean again.)

   Mean values are sensitive to outliers, which Honokāne is. Removing Honokāne will dramatically decrease the mean, and bring it closer to the median.

6. Make a histogram using the area of each ahupua'a.

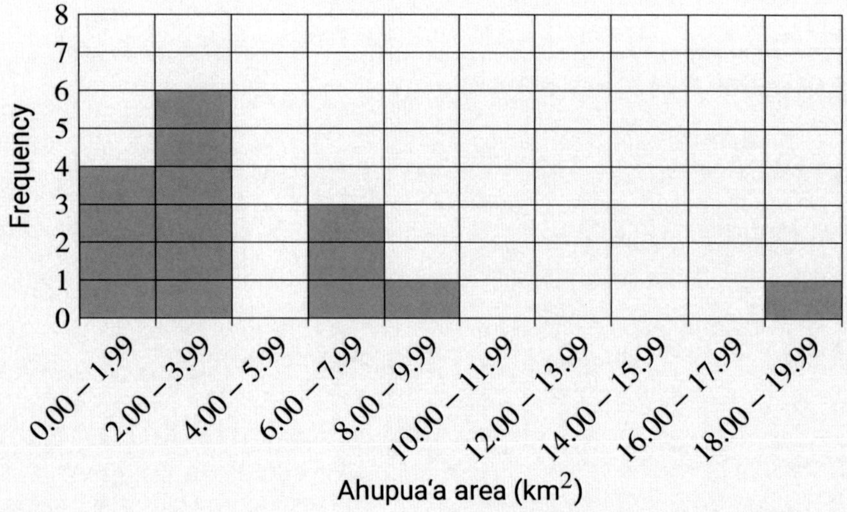

7. Which measure of center better describes the typical size of an ahupua'a in North Kohala–the median or the mean? Explain.

   Because the data has an outlier, the mean is less representative of the typical size of an ahupua'a. The median is not affected by outliers so it better describes the ahupua'a in North Kohala.

8. Choose an ahupua'a that you are familiar with but do NOT live in. Share something that you really like about that ahupua'a. Please share with your partner or in the online comment section.

Made in the USA
Las Vegas, NV
28 September 2024

95905224R00064